Energy Resources

Recent Titles in Contemporary Debates

ENERGY RESOURCES

Examining the Facts

Jerry A. McBeath

Contemporary Debates

An Imprint of ABC-CLIO, LLC
Santa Barbara, California • Denver, Colorado

Library of Congress Cataloging-in-Publication Data

Names: McBeath, Jerry A., author.
Title: Energy resources : examining the facts / Jerry A. McBeath.
Description: 1st edition. | Santa Barbara, California : ABC-CLIO, [2022] | Series: Contemporary debates | Includes bibliographical references and index.
Identifiers: LCCN 2022012411 (print) | LCCN 2022012412 (ebook) | ISBN 9781440869419 (cloth) | ISBN 9781440869426 (ebook)
Subjects: LCSH: Energy policy—United States. | Fossil fuels—United States—Environmental aspects. | Renewable energy sources—United States.
Classification: LCC HD9502.U52 M3916 2022 (print) | LCC HD9502.U52 (ebook) | DDC 333.790973—dc23/eng/20220405
LC record available at https://lccn.loc.gov/2022012411
LC ebook record available at https://lccn.loc.gov/2022012412

ISBN: 978-1-4408-6941-9 (print)
      978-1-4408-6942-6 (ebook)

26 25 24 23 22   1 2 3 4 5

This book is also available as an eBook.

ABC-CLIO
An Imprint of ABC-CLIO, LLC

ABC-CLIO, LLC
147 Castilian Drive
Santa Barbara, California 93117
www.abc-clio.com

This book is printed on acid-free paper ∞

Manufactured in the United States of America

Contents

---------------------❖❖❖---------------------

Acknowledgments

My interest in the broad subject of energy resources began about 40 years ago when undertaking a summer project for the Alaska Department of Natural Resources. The department's research office wanted a broad survey of oil and gas, coal and other hard rock minerals, as well as gravel and water resources of the Alaska North Slope. In the ensuing years, projects on Alaska energy and other natural resources as well as broader nationwide studies engaged me. My most recent book was *Big Oil in the United States*, published by Praeger in 2016.

Because this volume centers on important questions of energy and environmental policy, I sought advice from knowledgeable scholars. Most referred me to their works, and these are cited copiously in the volume. I want to thank specifically three friends who helped craft questions for a general audience: Dr. Milt Wiltse, Alaska Department of Natural Resources, Division of Geological and Geophysical Surveys (and former state geologist); Terrence C. Cole, emeritus professor of Public History, University of Alaska Fairbanks; and Mary A. Nordale, lawyer (former Alaska Commissioner of Revenue), also of Fairbanks.

Because writing takes time away from family, I want to thank my spouse Jenifer (professor of plant pathology and biotechnology, UAF), son Bowen (professor of social work and public policy, Portland State University), daughter Rowena (MD, partner, Philadelphia Hand to Shoulder Center and assistant professor, Thomas Jefferson University), and son-in-law David Zilkha, principal of the Zilkha family office. In different ways each provided valuable sounding boards for ideas and interesting diversions when needed.

At ABC-CLIO, my sincere thanks go to Kevin Hillstrom, Senior Acquisitions Editor, who patiently advised me throughout the project. I am grateful for the work Kristen Beach did in marketing the book, and for the assistance of Nicole Azze, Senior Production Editor, ABC-CLIO. Jitendra Kumar, Senior Project Manager, Westchester Publishing Services; and Gary Morris, Copyeditor, went through the manuscript painstakingly, and brought to light more typos and errors than can be mentioned quickly.

Of course, any remaining errors or omissions are my responsibility alone.

How to Use This Book

Energy Resources: Examining the Facts is part of ABC-CLIO's *Contemporary Debates* reference series. Each title in this series, which is intended for use by high school and undergraduate students as well as members of the general public, examines the veracity of controversial claims or beliefs surrounding a major political/cultural issue in the United States. The purpose of this series is to give readers a clear and unbiased understanding of current issues by informing them about falsehoods, half-truths, and misconceptions—and confirming the factual validity of other assertions—that have gained traction in America's political and cultural discourse. Ultimately, this series has been crafted to give readers the tools for a fuller understanding of controversial issues, policies, and laws that occupy center stage in American life and politics.

Each volume in this series identifies 30–40 questions swirling about the larger topic under discussion. These questions are examined in individualized entries, which are in turn arranged in broad subject chapters that cover certain aspects of the issue being examined, for example, history of concern about the issue, potential economic or social impact, or findings of latest scholarly research.

Each chapter features 4–10 individual entries. Each entry begins by stating an important and/or well-known **Question** about the issue being studied—for example, "Do some experts expect renewables to account for the majority of American energy consumption by 2045?" "Are partisan political considerations increasingly dictating how federal agencies and departments—including those related to the energy industry—operate and who leads them?" "Have oil and gas exploration and development reduced critical habitat of endangered species in the United States?"

The entry then provides a concise and objective one- or two-paragraph **Answer** to the featured question, followed by a more comprehensive, detailed explanation of **The Facts**. This latter portion of each entry uses quantifiable, evidence-based information from respected sources to fully address each question and provide readers with the information they need to be informed citizens. Importantly, entries will also acknowledge instances in which conflicting or incomplete data exists or legal judgments are contradictory. Finally, each entry concludes with a **Further Reading** section, providing users with information on other important and/or influential resources.

The ultimate purpose of every book in the *Contemporary Debates* series is to reject "false equivalence," in which demonstrably false beliefs or statements are given the same exposure and credence as the facts; to puncture myths that diminish our understanding of important policies and positions; to provide needed context for misleading statements and claims; and to confirm the factual accuracy of other assertions. In other words, volumes in this series are being crafted to clear the air surrounding some of the most contentious and misunderstood issues or our time—not just add another layer of obfuscation and uncertainty to the debate.

Introduction

From time immemorial, humans have relied on energy to survive—first to provide light, prepare food, keep warm, and protect against enemies; then to transport goods to markets (local and distant); and finally to power machines that ushered in an industrial revolution. In fact, human societies tend to be ranked by the *amount* and *type* of energy resources they use. Hunting and gathering societies once were regarded as primitive (and by some critics still are) because they put few demands on available sources of energy; for millennia they relied primarily on abundant wood products. Societies called "advanced" regarding economic growth today put heavy demands on multiple sources of energy, as no single source can supply all of the society's needs.

This increase in the amount of energy used by humans, and the growing diversification of energy types employed by human societies, have been accompanied by a large literature. The relationship between energy sources and economic development has been examined in such studies as those authored by Joan Robinson, Walt Rostow, and Robert Heilbroner. Many of these studies were undertaken in the early post-World War II era and were strongly influenced by changes in geopolitical forces (from multipolar to bipolar) stemming from that conflict. They often featured competition between rival nation-states for global hegemony as they sought monopolistic control over energy sources. Individual energy sources such as oil, natural gas, coal, and nuclear have been featured in historical and analytical studies, including influential scholarly volumes written by Daniel Yergin (including *The Prize* [2009], *The Quest* [2011], and *The New Map* [2020]).

The fall of the Berlin Wall in 1989 and dissolution of the Soviet Union in 1991 left the United States as the sole remaining superpower, which changed the nature of intellectual discussion. Threats to that status—for example, from a rising China and a renascent Russia—were recognized by researchers and scholars. In this era, however, analysts were increasingly critical of the historical waste of energy sources such as wood and oil, as well as the adverse environmental consequences of our steadily rising energy consumption. The concept of "sustainability" was first disseminated to a large audience in the 1987 Brundtland Report, a UN environmental study that became a measuring stick for assessing energy sources. In the last 20 years in particular, sustainability has become the central assessment of energy policy in the United States.

The primary focus of this volume is on the energy policies and priorities of three presidential administrations spanning two decades: George W. Bush (2001–2008), Barack Obama (2009–2016), Donald Trump (2017–2021), and the first 15 months of the Biden administration (2021–2022).

FOSSIL FUELS

Although the United States has relied on fossil fuels for energy longer than 100 years, specific fossil fuels have risen and fallen in popularity during that time. Coal, for example, accounted for nearly one-quarter of energy produced in the United States in the late 1990s. In 2019, though, it accounted for less than 16 percent of total electricity generated. More efficient drilling techniques reduced the cost of natural gas below those of coal and other generators of electric power at the same time that coal's negative environmental impacts attracted increased criticism. Although crude oil production declined from 1970 to 2008, the increase of exploration for shale oil in 2009 and thereafter reversed this trend.

Supply and demand. The laws of supply and demand strongly influence the price of fossil fuels in the United States. *Ceteris paribus* (other things being equal), quantity demanded is inversely related to price; quantity supplied is directly related to price. Although the American market is one of the world's largest, the high degree of transportation connectivity means that domestic differences in supply and demand have an indirect impact on regional and local prices. Global conditions of supply and demand, especially regarding petroleum, have a stronger influence over price. However, two conditions buffer the impact of changes in supply and demand.

First, in the early years of oil and gas production, multinational corporations exercised enormous power over the industry as a whole. The weakening and eventual collapse of most colonial powers after the First and Second World Wars, however, weakened the position of multinational energy corporations with operations in these newly independent nations. Also, the postwar development of nationalist movements challenged the dominance of Western institutions and forces over fossil fuel exploitation (Engler, 1961).

Second, in 1969, leaders of Middle East oil-producing countries (especially Saudi Arabia, Iran, Iraq, and Libya) joined with other developing nations to form the Organization of the Petroleum Exporting Countries (OPEC). This organization in economic terms was a *cartel*—formed for the purpose of increasing oil prices (to the benefit of producing nations) by limiting production. OPEC demonstrated its power over oil markets through two oil shocks. In the first, following the Yom Kippur War of October 1973 (in which Israel defeated Arab armies), OPEC members established an oil embargo against the United States and the Netherlands. The sharp falloff of oil supplies in the United States resulted in a quadrupling of prices and signs of scarcity such as long lines outside gas stations.

The second oil shock occurred following the onset of war between Iran and Iraq, two major oil-producing states. This war, which raged from 1980 to 1988, stopped production in many oil fields, leading to substantial reductions in supply and further increases in the price of oil. Oil is an essential commodity of the modern world, critical for transportation as well as industrial production. While a drop in the supply of apples is not immediately or necessarily registered by a rise in apple prices (because one can eat alternate fruits or go without), a rise in oil prices is accompanied by no countervailing fall in demand. This is what economists call *inelastic demand* and means that the market is not able to quickly adjust to volatility.

The power of OPEC has waxed and waned since the 1970s. The development of new fields outside OPEC's control—in the Alaska North Slope, the province of Alberta in western Canada, the British/Norwegian discoveries in the North Sea, and in Mexico—reduced OPEC's influence over the market prices of oil. Yet fossil fuels are not renewable, and when production from established fields in Europe and North America declined, OPEC's advantages increased. Even some of the fields considered new in the late 1960s and 1970s have become depleted.

Preview of chapter 1. Fossil fuels supplied most energy in the U.S. until the twenty-first century, but the transition to clean energy is on the way. The supply of oil, natural gas, and coal will not be exhausted soon, as new reserves are located and new technology (hydraulic fracturing and horizontal drilling) are employed. The shale revolution returned the United States to a condition of virtual energy independence in 2018. However, demand for fossil fuels has waned because of volatile global prices and increased awareness of adverse short- and long-term (climate warming) environmental effects.

Huge corporations dominate exploration, production, and distribution of fossil fuels, but market structures have evolved. Multinational oil, gas, and mining firms are divesting themselves of some fossil fuel reserves, and in some cases diversifying into renewables. More government grants, contracts, and tax subsidies have supported fossil fuels, but this picture too is changing. During the 2019–2021 period, geopolitical competition slowed economic growth, and the COVID-19 crisis brought natural resource price spikes. Yet the once unthinkable—electric vehicles (EVs) rising to account for one-third of new vehicle purchases by 2030—has become plausible.

Preview of chapter 2. Some renewables, such as hydro- and nuclear power, have been significant parts of the U.S. energy and electricity systems for more than a half-century. They were built in an era of big projects, and now their infrastructure is aging and repairs will be costly. The best future for hydro/nuclear lies in construction of mini-reactors and small-scale hydro plants. Other renewables, such as solar and wind power, have become practical electricity sources just in the last few decades. Their prices have become more competitive, and technological breakthroughs have increased capacity of battery (and other forms of) storage. Obstacles to rapid expansion of solar/wind power include problems of infrastructure, transportation, supply chain shortages (of critical metals such as lithium and cobalt), and the uncertain investment climate, as well as Republican-led states that have generally been more supportive of fossil fuel drilling and mining than of solar/wind power facilities.

Preview of chapter 3. Minor renewables (biomass, corn-based ethanol, biofuel/gases, geothermal and tidal/wave power) produce less than one-twentieth of the energy consumed in the United States today. Yet in a country increasingly conscious of the damages of pollution and greenhouse gas emissions to air/land/water and the broader ecosystem, they offer great promise. Bringing these resources to market will take very large investments from the private sector and government as well as technological fixes.

Preview of chapter 4. The U.S. power grid (electricity system) faces increasing challenges to its reliability, capability, sustainability, and affordability.

For example, grid blackouts in the United States are more frequent than in grids of other developed nations, and just within the last few years major blackouts occurred in California and Texas. Most nations have a national grid, but the U.S. grid features a patchwork of power plants, transformers, high-voltage transmission lines, and distribution lines bundled into regions.

Meanwhile, threats to America's energy grid continue to grow in a variety of forms, including increased vulnerability to electromagnetic pulses and geomagnetic shock waves; extreme weather events; and cyberattacks.

Preview of chapter 5. The period covered most carefully in this volume (2000 to 2022) has been characterized by extreme partisan polarization and political gridlock, making it extremely difficult for the United States to maintain a coherent, consistent, and effective energy policy.

Critiques of what could be called the "energy bureaucracy"—the regulations and policies of the EPA and the departments of Energy and Interior—have intensified, with conservative lawmakers at the local, state, and federal levels all laboring to reduce the influence of these agencies over industry—or to hand the reins of these departments and agencies over to pro-industry leadership.

Preview of chapter 6. Plans to seek, develop, and produce energy resources (claimed to be critical for economic growth in the United States) often conflict with environmental values. In polarized eras, compromises are immensely difficult to forge. Instead, one side wins, the other loses. The size of the win or loss varies by presidential administration and its coalition of support.

Extensive policy and public opinion research demonstrate the divide between supporters and opponents of taking action to address climate change. Trade-offs concerning other major environmental legislation show similar divisions. The chapter considers debates over changes in air quality (measured under terms of the Clean Air Act and its amendments), regulation of carbon dioxide emissions via Corporate Average Fuel Economy (CAFE) standards, legal and political struggles over the Clean Water Act, the Endangered Species Act, and other environmental and conservation laws and regulations impacting the energy industry.

Preview of chapter 7. The energy transition was delayed by the global economic slowdown and the international health crisis of the COVID-19 pandemic, with a pronounced decline from 2019 to mid-2021. Like much of the rest of the world, the United States entered 2022 on a slow but steady path of economic recovery. Even so, a national consensus in perspective and action on energy and environmental policy seems unlikely in the near term.

FURTHER READING

E&E News. "Big oil is selling dirty assets, but they aren't going away," October 9, 2019.

Economist. "Bigger oil," February 9, 2019.

Engler, Robert. *The Politics of Oil: Private Power & Democratic Directions.* Chicago: University of Chicago Press, 1961.

Heilbroner, Robert and William Milberg. *The Making of Economic Society,* 13th ed. New York: Pearson Higher Education, 2011.

Robinson, Joan. *The Accumulation of Capital.* London: Macmillan, 1956.

Rostow, Walt. *Politics and the Stages of Growth.* London: Cambridge University Press, 1971.

U.S. Department of Energy, Energy Information Administration (EIA). "U.S. energy facts explained," May 14, 2021. https://www.eia.gov/energy explained/us-energy-facts

Yergin, Daniel. *The Prize: The Epic Quest for Oil, Money & Power.* New York: Free Press, 2009.

Yergin, Daniel. *The Quest: Energy, Security, and the Remaking of the Modern World.* New York: Penguin, 2011.

Yergin, Daniel. *The New Map: Energy, Climate, and the Clash of Nations.* New York: Penguin, 2020.

1

❖❖❖

Fossil Fuels

This chapter examines fossil fuels from a variety of perspectives, including the extent to which fossil fuels have been exhausted. One question asks whether market concentration among oil and gas producers has increased in the United States, and if so, whether this has had an effect on price. Given the history of instability in oil supplies and shocks, consideration is given to whether U.S. dependence on Middle Eastern oil has been reduced. A different way to ask this question is to explore whether the announced Trump administration policy of "energy dominance" through fossil fuels has been effective.

Both domestic and global influences affect fossil fuel exploration and development in the United States. The shale oil and gas revolution of the early twenty-first century can be interpreted as driven by either foreign or domestic demand. For one of the fossil fuels (coal), however, domestic forces are dominant, and the question in this regard asks whether pro-coal administrative policies (an example of industrial policy, of the government telling the market how to behave) have had a greater effect than market forces on actions of utilities.

One aspect in the discussion of renewable forms of energy such as solar and wind is that they must compete with fossil fuels that have historically benefitted from government subsidies. This raises the question whether, overall, subsidies are in fact greater for some forms of energy than others. Another facet of the fossil fuel versus renewable debate concerns electric vehicles (EVs) ranging from full-time electric cars to hybrids such as the Prius to Tesla's self-driving vehicles. Has development of EVs seriously challenged the market share of gasoline-only vehicles? Could they in the future? And where do EVs figure in the calculations of Big Oil?

The chapter concludes with questions taking us full circle. Are the press attention and public discussion about renewables warranted? Do they really have the potential to transform America's energy profile in the years to come?

Q1. IS THE UNITED STATES LIKELY TO RUN OUT OF FOSSIL FUELS BY 2045?

Answer: No, the consensus of energy experts is that domestic supplies of U.S. fossil fuels and particularly oil and natural gas will not be exhausted in the next generation.

The Facts: Discussion of domestic supplies of fossil fuels typically orbits around five elements or concepts: peak oil, increased demand, technology, oil field revivals, and energy security.

Peak oil. The term "peak oil" is used to denote the idea that extraction of oil will reach a peak after which production will decline permanently. Kenneth Deffeyes, a Princeton University geoscientist, discussed the concept in his 2001 book *Hubbert's Peak: The Impending World Oil Shortage.* He recalled the prediction of geologist M. King Hubbert that oil production in the United States would peak in the 1970s, followed by a period of great instability. Hubbert's belief had many adherents, but the date was pushed back. (A related bet between business professor Julian Simon and biologist Paul Ehrlich concerned whether raw materials' prices [five metals] would rise or decline over the course of the 1980s. Simon said they would decline and won the bet [Tierney, 1990].) It is clear that predictions of exhausting oil and gas supplies by certain dates, or of steep and persistent high oil and gas prices, have not been borne out yet. They continue to be made, however. For example, the analytics firm Rystad Energy estimated that peak oil demand could arrive within 10 years of the start of COVID-19 (Willson, 2020). Still, energy scholar Daniel Yergin believes that it is premature to generalize based on the COVID-19 experience, and opines that demand will return to previous levels by 2022–23 (Yergin, 2020). Certainly, as of early 2022, these two fossil fuels on which most nation-states depend continue to play central roles in their economies and cultures, and have not yet been priced beyond the means of most consumers. (The term "peak oil" was used by Shell when it indicated that it would dial down production 1 to 2 percent annually, with production of traditional fuels falling to 55 percent by 2040 [Sobczyk, 2021]).

Increased demand. Among volatility represented in fossil fuel gluts and scarcities, analytical reports suggest that to satisfy needs of the near term, drilling for oil and gas must double beyond current levels. This was the finding of the International Energy Agency (IEA) in its *World Energy Outlook 2018* (IEA, 2018). Researchers urged increased production from both conventional oil projects and shale oil and gas projects, pointing to needs of economically developed nations as well as the emerging economies of China and India. The forecast stated that government policy would be the primary driver of energy supply and consumption patterns, strongly influencing investments. In mid-2021, the Biden administration called for continued use of natural gas, as natural gas and oil prices skyrocketed, in a boom for fossil fuels including coal (Clark, 2021). And at the Russian invasion of the Ukraine in February 2022, President Biden urged U.S. oil/gas companies to increase production to assist European NATO allies dependent on Russia for 40 percent of their natural gas supplies (Soraghan & Richards, 2022).

Technology. A third factor regarding fossil fuel "constancy" is advances and investments in new technologies. The shale revolution (combining hydraulic fracturing or fracking [HF] with horizontal drilling to squeeze oil and natural gas from shale, which vastly increased U.S. energy production) is made possible by technological innovation in the oil and gas industries. It emphasizes horizontal (as opposed to traditional vertical) drilling, and expanding drilling pipes to far greater distances underground. This method of extraction has reduced the size of drilling pads greatly, while increasing the use of water, sand, and chemical drilling muds, all of which have enabled the industry to extract oil and gas out of once impenetrable rock. Offshore, technological changes have allowed producers to increase the depth of drilling into the ocean floor, expand causeways miles into the nearshore, and construct gravel islands, new terminals, and other spaces for transporting resources to markets.

Oil field revivals. The United States has experienced a dramatic upsurge in production from fields once thought to be nearly depleted—as well as rising production from basins not yet fully explored. Revived fields include Prudhoe Bay and Kuparuk on the Alaska North Slope and areas of Pennsylvania where commercially feasible oil was first discovered in 1859. Improved technology has enabled drillers to operate in oil- and gas-rich regions of most U.S. states in the early twenty-first century. The largest fields worked on in the last decade include the west Texas Permian Basin and the Marcellus Shale northern Appalachian Basin. Notwithstanding the environmental and business risks of drilling in the Outer Continental Shelf (OCS), interest in drilling offshore waters has grown as well (Brugger, 2019). Because offshore oil/gas constitutes about 16 percent of total U.S. production, corporations have paid attention to federal leasing schedules (managed by the Bureau of Ocean and Energy Management, BOEM), and placed bids on the most attractive sites, the majority of which are in the Gulf of Mexico region. In addition, large oil/gas discoveries have been made in areas not previously targeted, such as 88 Energy Ltd.'s discovery in the Nanushuk Reservoir (between Willow and Umiat oil fields of the National Petroleum Reserve-Alaska) (Richards, 2021).

Energy security. The fifth element of discussion is the role that energy independence has played in American national policy from the Carter administration to the Biden presidency. For 70 years after 1949, U.S. annual imports of oil and gas exceeded exports. Although most of the imports came from Canada and Mexico, roughly a third were from Organization of the Petroleum Exporting Countries (OPEC) members, including nations in the politically volatile Middle East. In late 2019, though, *Forbes* declared that the United States had become a net exporter of oil and petroleum products (Cohen, 2019). (This situation was at least temporarily reversed when the COVID-19 pandemic struck in early 2020 and the United States returned to being a net importer of crude oil and petroleum products [EIA, 2020].) A confluence of forces—including the shale revolution, changes of administrations (and their policy emphases), growth of several industrial sectors at home, and economic problems abroad—figured in this development, and it may be a brief episode given that the United States remains subject to supply dislocations and demand changes linked to economic growth.

The oil glut after early January 2020 led to work slowdowns and layoffs, and ultimately to broad-ranging unemployment of oil field workers and allied service firms (Anchondo, 2020; Lee & Anchondo, 2020; Richards, 2020). Projects were put on hold (Anchondo & Lee, 2020), and drilling for oil and gas fell to the lowest level since 1948 (Soraghan, 2020). Reduction in demand also spurred by the COVID-19 crisis resulted in American oil and gas production exceeding consumption for the first time since 1957 (Clark, 2020). However, by early 2022, these conditions had changed, and the Energy Information Administration (EIA) predicted that crude oil production would surpass pre-pandemic levels, reaching "the highest annual average U.S. crude oil production on record."

FURTHER READING

Anchondo, Carlos. "Exxon warns of $3.1B hit as coronavirus crash worsens," *E&E News*, July 6, 2020.

Anchondo, Carlos and Mike Lee. "'Tough decisions,' Coronavirus spurs project delays, layoffs," *E&E News*, March 20, 2020.

Brugger, Kelsey. "Despite risk, interest in offshore waters grows," *E&E News*, February 15, 2019.

Clark, Lesley. "U.S. production beats consumption for first time since 1957," *E&E News*, April 29, 2020.

Clark, Lesley. "Energy crisis tests Biden's clean electricity agenda," *E&E News*, October 14, 2021.

Cohen, Ariel. "Making history: U.S. exports more petroleum than it imports in September and October," *Forbes*, November 26, 2019.

Deffeyes, Kenneth S. *Hubbert's Peak: The Impending World Oil Shortage*. Princeton, NJ: Princeton University Press, 2009.

International Energy Agency (IEA). *World Energy Outlook 2018*. December 13, 2018.

Lee, Mike and Carlos Anchondo. "Halliburton revenue drops $1.8B as rival files for bankruptcy," *E&E News*, July 21, 2020.

Richards, Heather. "BP cuts 10,000 jobs: We have to spend less money," *E&E News*, June 8, 2020.

Richards, Heather. "Huge Arctic oil find makes waves," *E&E News*, September 1, 2021.

Sobczyk, Nick. "Shell says it hit peak oil, will go carbon neutral," *E&E News*, February 11, 2021.

Soraghan, Mike. "Oil, gas drilling hits lowest level since 1949," *E&E News*, June 1, 2020.

Soraghan, Mike and Heather Richards. "Biden to oil industry after Russian ban: Drill," *Energywire*, March 9, 2022.

Tierney, John. "Betting on the planet," *New York Times*, December 2, 1990. https://www.nytimes.com/1990/12/02/magazine/betting-on-the-planet.html

U.S. Department of Energy, Energy Information Administration (EIA). "EIA expects the United States will return to being a net importer of crude oil and petroleum products," April 1, 2020 (press release).

U.S. Department of Energy, Energy Information Administration (EIA). "EIA expects annual U.S. crude oil production to surpass pre-pandemic levels in 2023," January 11, 2022.

Willson, Miranda. "Peak oil demand could come sooner than expected—report," *E&E News*, June 22, 2020.

Yergin, Daniel. *The New Map: Energy, Climate, and the Clash of Nations*. New York: Penguin Press, 2020.

Q2. HAS MARKET CONCENTRATION INCREASED AMONG U.S. OIL AND GAS PRODUCERS?

Answer: Yes, there has been a modest rise in concentration, primarily to increase efficiencies at a time of very low oil and gas prices, to reduce vulnerability of firms by providing access to more capital (to meet operating expenses) and to increase strength through cartelization.

The Facts: All degrees of concentration have been manifest in the history of oil and gas production in the United States—ranging from individual producers to monopolies—but the norm has been a relatively concentrated pattern of control. This industry environment has resulted in higher prices than one would find in a completely free market.

In the era of very rapid economic development following the Civil War (called the "Gilded Age," from the 1870s to the late 1890s), concentration of capital arose in several sectors, such as transportation, steel, and oil/gas. Popular lore featured "robber barons" of the railroads, and John D. Rockefeller, whose corporation Standard Oil had a monopoly over exploration, production, and distribution of oil in the United States (Chernow, 1998). In response to control over entire economic sectors—industries that were also central to the nation's industrial and economic expansion—Progressive forces in Congress passed the Interstate Commerce Act of 1887, which established the Interstate Commerce Commission, the first federal authority to regulate big business. This was followed by adoption of the Sherman Anti-Trust Act of 1890, passed explicitly to break up business monopolies. The trusts successfully fought the Sherman Act in the courts, but Theodore Roosevelt (president from 1901 to 1909) gained the reputation as a trustbuster for his attacks on the Northeast transportation monopoly and also Rockefeller and the Standard Oil trust. It was not until the succeeding Taft administration that the Supreme Court found Standard to be in violation of the Sherman Act and split the monopoly into 34 separate companies. These included Standard Oil of New Jersey (later renamed Esso, then Exxon, and now ExxonMobil), Standard Oil Company of New York (Socony, later Mobil, also now part of Exxon Mobil), and Standard Oil of California (now Chevron) (Engler, 1961; McBeath, 2016).

Breaking up Standard Oil did not end the concentration of political and economic power, for a small number of large firms continued to dominate exploration, production, and distribution of oil and gas. By the 1920s, seven huge corporations dominated the global oil trade: three American offspring of Standard Oil (Chevron, Exxon, and Mobil), two Texas firms (Gulf Oil and Texaco), the British Anglo-Iranian Oil Company (now British Petroleum [BP]), and one Anglo-Dutch (Royal Dutch Shell [Sampson, 1991]). Named the Seven Sisters to indicate they competed and cooperated much like the Seven Sisters of Greek mythology, each was a multinational corporation and fully integrated. Their global dominance was challenged by post-World War II nationalization of oil deposits in the Middle East, Africa, and Latin America and the formation of the Organization of the Petroleum Exporting Countries (OPEC) in 1960.

Domestically, the New Deal administration of Franklin Delano Roosevelt (FDR) tempered antitrust sentiments of the Progressives as part of economic recovery from the Great Depression, and made attempts similar to cartelization (essentially, price-fixing of large firms, but allowed by government). In fact, some political economists call the period from 1933 to 1972 "crude oil cartelization," represented by the Texas Railroad Commission's restraints on petroleum production (Childs, 2005; Libecap, 1989).

Research on mergers and acquisitions. Beginning in the 1990s, the petroleum industry underwent a series of mergers, acquisitions, and joint ventures, which in the view of critics threatened to raise prices greatly for consumers and substantially reduce competitiveness in this sector. One of the largest was Exxon's 1999 acquisition of Mobil, a former competitor, to create the giant ExxonMobil corporation (Yergin, 2011). Members of the House and Senate asked the U.S. Government Accountability Office (GAO), an instrumentality of the Congress, to report on this development.

The GAO's May 2004 report indicated that there had been 2,600 mergers since the 1990s, a relatively large number, most frequently among firms involved in energy exploration and production. Interviews with company officials suggested that a leading motivation for these mergers and acquisitions was increased efficiency and cost savings, when energy prices were low. Because economists find a tendency for mergers to reduce the ability of new firms to enter markets, thereby reducing competition (and increasing prices), the GAO auditors reviewed the evidence. Their econometric analysis concluded that increased market concentration had resulted in higher gas prices from the mid-1990s to 2000, with differences among retail and wholesale commodities (GAO, 2004). Another result of growth in concentration was increased spending on industry lobbying (Lee, 2021).

The "energy transition." Energy prices were low in the 1990s; they climbed in the early twenty-first century, reaching $147/barrel for oil in 2008, and then dropped about 60 percent from that high point to relatively low levels toward the end of the next decade. This time period corresponded with the onset of the Great Recession in the United States, but also with growing national and international attention to adverse environmental consequences of fossil fuel production and consumption. Environmental protests became frequent, and fossil fuels were easy targets.

The global focus on greenhouse gases as major human causes of climate warming (and particularly carbon dioxide and methane emissions from fossil fuel combustion) led to international activity in the Kyoto Protocol (1997) and Paris Climate Accords (2017). Many corporation leaders became defensive about the fossil fuels their firms produced and more open-minded about investing in renewables, especially wind and solar.

Lower energy prices encouraged a nearly universal attempt to make oil and gas exploration and production more efficient, with estimates of an average 10 percent reduction in costs likely across-the-board. New technologies were thought to be harbingers of positive change in this regard, with digitalization of production—replacing workers with computers and increasing automation of processes—leading to reduced employment in many industry sectors. Successful firms shed workers despite cost savings, while other firms that could not adapt to the fast-changing environment filed for bankruptcy. These business failures further diminished the size of the industry workforce (Lee, 2020).

Concentration in the current energy market. Three changes occurred in the energy sector. Firms pursued merger and acquisition (M&A) strategies to reduce costs, often with unexpected outcomes (wins became losses when conditions changed); other firms sold off reserves and in several cases found that the assets declined in value when oil prices crashed; finally, a number of fossil fuels firms increased investment in renewables. Examples follow for each type of change.

In 2019, Occidental competed with Chevron to acquire Anadarko, which had prime oil and gas assets in the United States. Occidental CEO Vicki Hollub offered a high price of $55 billion to seal the deal, expecting to be able to sell off other holdings before large debt payments came due. Then oil prices crashed to $20/barrel, Occidental's stock fell 75 percent, and there was no market for the assets it sought to sell. Credit agencies such as Moody's relegated its corporate bonds to junk status (Osborne, 2020). Meanwhile, lower prices gave Big Oil advantages that small, independent oil and gas companies did not possess. In the Permian Basin, for example, Exxon and Chevron (both integrated corporations owning not only land but pipelines and refineries) were less vulnerable to shale oil and gas price declines than their smaller counterparts and could still make good profits at lower prices (Lee, 2018).

Multinationals were not averse to selling off reserves whose value had declined. In 2017, for example, ExxonMobil wrote down (deliberately reduced the book value of) 4.8 billion barrels of reserves, because it was not feasible to exploit them at low oil prices. This increased the corporation's flexibility, and it was able to develop oil in new areas abroad, such as Guyana (*Economist*, 2019). British Petroleum (BP), which had held the larger interest in reserves in Prudhoe Bay and the associated Trans-Alaska Pipeline System (TAPS) sold its interest in what it deemed to be an old and depleted field to Hilcorp in 2019. The latter firm is mostly privately held by billionaire owner David Hildebrand, who has a reputation for turning profits from declining oil and gas fields and had the cash to pay BP $5.6 billion to conclude the deal. As a publicly traded corporation with shareholders sensitive to climate warming protests, BP had an incentive to shed "dirty" assets, while Hilcorp faced little such pressure and could promise Alaska officials it had a stronger incentive to boost production than BP (*E&E News*, 2019). ConocoPhillips, BP's major partner in its northern Alaska oil field operations, has announced plans to sell off assets too.

The third set of examples pertains to fossil fuel firms now investing in renewables, a different direction of diversification than used previously (when firms invested in pipelines, refineries, and gas stations). Chevron is one U.S. company that has made significant investments in wind power (see also Q8), and it spent $3.15 billion to buy the nation's largest biodiesel producer (Lee, 2022). To the present, however, energy corporations head-quartered in the United States have been less aggressive in renewables investing than giants like BP and Shell who are headquartered in Europe, where growth in renewables is stronger. In fact, BP sought FERC approval to sell power in five American states, denoting its shift to renewables (Willson & Anchondo, 2021). Of possibly greater significance is the decision of large corporations to invest in non-oil businesses, such as electrical vehicle charging stations and chemicals and plastics manufacturing, which increases the diversification of their portfolios.

The first large deal since the crash of oil prices was Chevron's purchase of Noble Energy for around $5 billion in October 2020. To some observers this announcement signaled the potential start of a sweeping consolidation in the American oil industry. As the vice-chair of energy consultant Deloitte, Duane Dickson, remarked: "In a downturn like this, the strong get stronger and the weaker players try to survive as best they can, and some will be bought" (Krauss, 2020). Chevron itself had suffered losses, but its leaders expressed confidence in industry recovery soon.

Finally, the increased concentration of oil and gas firms has been broadened into eight sea industries, which generate more than $1 trillion in revenue annually. A recent study found that just 100 companies controlled the lion's share of offshore oil and gas, offshore wind, marine equipment and construction, seafood production and processing, container shipping, shipbuilding and repair, cruise tourism, and port activities (Yurk, 2021).

FURTHER READING

Chernow, Ron. *Titan: The Life of John D. Rockefeller, Sr.* New York: Random House, 1998.

Childs, William R. *The Texas Railroad Commission: Understanding Regulations in America to the Mid-Twentieth Century.* College Station, Texas: A&M University Press, 2005.

E&E News. "Big oil is selling dirty assets, but they aren't going away," October 9, 2019.

Economist. "Bigger oil," February 9, 2019.

Engler, Robert. *The Politics of Oil: A Study of Private Power & Democratic Directions.* Chicago: University of Chicago Press, 1961.

Krauss, Clifford. "Chevron deal for oil and gas fields may set off new wave," *New York Times,* July 20, 2020.

Lee, Mike. "Supermajors high on gas exports," *E&E News,* February 4, 2019.

Lee, Mike. "'Google of oil and gas' files for bankruptcy," *E&E News,* June 29, 2020.

Lee, Mike. "Report probes 'dangerous feedback loop' of oil lobbying," *E&E News,* August 26, 2021.

Lee, Mike. "Chevron buys renewable fuel giant in record $3B deal," *Energywire,* March 1, 2022.

Lee, Mike and Saqib Rahin. "Tax law boosts oil majors during tepid earnings season," *Energywire,* February 5, 2018.

Libecap, Gary D. "The political economy of crude oil cartelization in the United States, 1933–1972," *Journal of Economic History,* Vol. 49, no. 4 (December 1989): 833–55.

McBeath, Jerry A. *Big Oil in the United States: Industry Influence on Institutions, Policy, and Politics.* Santa Barbara, CA: Praeger, 2016.

Osborne, James. "With $39 billion in debt, can Houston-based Occidental Petroleum survive the oil crash?" *Houston Chronicle,* March 20, 2020.

Sampson, Anthony. *The Seven Sisters.* New York: Bantam, 1991.

Seattle Times. "BP, Hilcorp say deal involving Alaska assets, interests done," December 18, 2020.

U.S. Congress, Government Accountability Office (GAO). "Energy markets: Effects of mergers and market concentration in the U.S. petroleum industry," May 2004. https://www.gao.gov/products/gao-04-96

Willson, Miranda and Carlos Anchondo. "In shift to renewables, BP seeks FERC approval to sell power," *E&E News,* April 27, 2021.

Yergin, Daniel. *The Quest: Energy, Security, and the Remaking of the Modern World.* New York: Penguin, 2011.

Yurk, Valerie. "100 companies dominate sea industries like offshore energy," *E&E News,* January 15, 2021.

Q3. IS THE UNITED STATES DEPENDENT ON MIDDLE EASTERN OIL?

Answer: No, the United States is mostly self-sufficient in oil, but it does have a "thinning reliance" on Middle Eastern crude oil for American refineries.

The Facts: From 2018 to early 2020, the United States produced more oil than it consumed (EIA, 2019), and was again independent. A more nuanced discussion considers briefly the history of U.S. oil imports, the search for stability to avoid oil price volatility and its disruptive effects, responses of domestic as well as international forces to the energy trade, and the oil glut and price crash of early 2020.

Around the turn of the twentieth century, the United States was a global oil giant with large surpluses, and its multinational corporations sold a range of petroleum products in most parts of the world. National policies favored protection of domestic oil, including import bans, embargoes against foreign powers with hostile intentions (such as Japan before the "Pacific War," the theatre of World War II that encompassed Asia and much of the Pacific and Indian oceans), and oil aid to allies in two world wars. Beginning in 1949, however, the United States began to import more oil than it produced. This remained the case until 2018, when the shale revolution (combining hydraulic fracturing with horizontal drilling to squeeze oil and natural gas from shale) took effect and America became a net exporter of oil once again.

Search for stability. During the seven decades when domestic oil production sagged, most observers called the United States an oil-dependent nation. Oil, in their view, was a trade good that was unlike automobiles, cheese, and corn, all of which could be allocated according to market rules of supply and demand. Oil, however, was so critical to overall economic health that many observers believed federal intervention in the oil market was justified. After the oil shortages of the 1970s, during which some Americans waited hours in long lines at gas stations to fill up their tanks, presidents uniformly sought "U.S. energy independence." Critics, on the other hand, questioned whether independence from oil imports was realistic (Cordesman, 2013). And when oil in other nations became abundant and cheap again, it seemed wise policy to deplete their resources first and conserve domestic reserves.

Domestically, policies protective of domestic oil production were slow to change. Bans on importing oil protected the higher prices of U.S. oil against foreign competition, until restrictions were lifted in April 1973. Unfortunately for American consumers, shortly after gates opened to a flood of foreign oil, OPEC took actions in the most significant energy crisis of 1973–74. (In 1960, nations with globally significant oil/gas deposits and resentful of the Seven Sisters and nations in which they were headquartered formed a cartel, called the Organization of the Petroleum Exporting Countries [OPEC]). In October 1973, OPEC's Arab state members (with a controlling interest in the cartel) declared they would cut oil production and embargo oil sales to certain countries (especially the United States) in protest of U.S. support for Israel in the Yom Kippur War. This created a shortage in oil supply, and prices of gasoline nearly quadrupled to more than $12/barrel.

Although OPEC had influence, it controlled only 56 percent of the international oil market. At the outset of the Iranian revolution in 1979 (which replaced the secular Shah, an American ally, with the religious leader Ayatollah Khomeini), oil production from a once stable source of U.S. imports stopped. Following this revolution, Iraq's leader Saddam Hussein invaded Iran, and the ensuing war brought about a dramatic decline in oil production of both countries. By July 1980, the oil market price was $30, more than double the $13 market price of the previous year. By this time, however, new sources of oil (from Alaska, the Gulf of Mexico, Canada, the North Sea, and Siberia) had stabilized oil supplies (Yergin, 2011).

Import bans were lifted only after the oil shocks of the 1970s; restrictions on exporting U.S. oil remained in effect until 2015. Moderating the disruptive impacts of OPEC price hikes was a constant policy of prioritizing oil imports from Canada and Mexico, as well as imports from Latin American nations such as Ecuador and Venezuela.

Continued small imports. However, even after American petroleum exports surged in 2019, the United States continued to import about 5 percent of the oil it consumed (EIA, 2021). The major sources of imports have been Canada, Saudi Arabia, Mexico, Venezuela, and Iraq. The amount invested in oil and gas infrastructure suggests that some imports will continue into the future. Most refineries are equipped to process heavy and "sour" oil (which contains sulfur compounds), whereas the new shale oil resources that have increased U.S. oil supplies are light and "sweet," making them undesirable for refiners. As a result, much of the oil extracted from America's shale fields is exported (Rapier, 2018).

Internationally, the United States stands again as the world's largest producer of oil, but the global production environment is both more competitive and more volatile than it was a century ago. As a nation that is seen as more economically and politically stable than many other big oil producers, the United States seems likely in the future to continue to safeguard producers such as the Gulf Cooperation Council members Saudi Arabia and Iraq—which agree with many U.S. foreign policy goals, chief among which is constraining Iran. Even in the populist Trump administration, which acerbically challenged America's old alliances (e.g., NATO), there were intense efforts to have allies rejoin sanctions on Iran. (In 2015, the Obama administration and European allies agreed to remove sanctions on Iran's trade in oil in exchange for major restrictions on its nuclear energy program. In 2018, President Trump withdrew from the deal, complaining that Iran had not met requirements, and reimposed sanctions.)

The increased complexity of the contemporary world order—with an economically strong and rising power (China) and a nuclear rival (Russia) having revived imperialist ambitions—makes oil a potentially useful instrument of U.S. foreign policy. Yet other organized forces, particularly OPEC, challenge U.S. influence. The United States alone cannot set the international oil market price, even if it is nearly self-sufficient in oil and gas resources, notwithstanding U.S. petroleum wealth in 2022. In other words, there are limits to the extent to which American energy independence can change foreign policy outcomes.

In early 2020, when Iranian missiles struck a joint U.S.-Iraqi base, President Trump said the United States would impose punishing economic sanctions on Iran. He further noted: "We are now the No. 1 producer of oil and natural gas anywhere in the world. We are independent, and we do not need Middle East oil" (Dillon, 2020b). Several observers, including *E&E News*, set the record straight, noting that Middle East oil made up less than 10 percent of imports and "America's thinning reliance on (it) . . . isn't about to reverse course" (Dillon, 2020b).

Oil crashes compounded by the COVID-19 pandemic. The price war in late 2019/early 2020 between Saudi Arabia and Russia for control of the global oil market pushed prices to extremely low levels. Because American shale producers had higher costs than oil producers in most Middle Eastern countries, they were hard hit. HIS Markit consultant Jim Burkhard remarked: "If this situation persists amidst a recession, it points to the possible buildup of the most extreme global oil supply surplus ever recorded" (Lee, 2020). Conditions improved somewhat after U.S. oil production seemed to bottom out in mid-2020 (Adelman, 2020), but U.S. oil supplies were not expected to return to pre-COVID-19 levels until 2021 or 2022 (EIA, 2020; Yergin, 2020).

Eruption of the COVID-19 pandemic in early 2020 and its cascading effects seriously compounded the downward spiral of oil prices, which dropped below zero briefly at the end of the first quarter. "Negative" prices for oil means for some producers it was cheaper in the long term to take a significant loss on future oil contracts instead of finding a place to store oil or shutting down production (Tobben, 2020). Drilling rigs sat idle, layoffs spread, and the number of storage places for surplus oil vanished quickly (Lee et al., 2020). Low prices in the U.S. market meant limited incentives for new exploration or for ramping up existing production (Richards, 2020). As noted in Q1, the situation had changed by mid-2021. Ironically, the oil supply crunch following Russia's invasion of the Ukraine in early 2022 prompted President Biden to reach out to countries such as Saudi Arabia that Democratic presidents have snubbed (Associated Press, 2022).

FURTHER READING

Adelman, Bob. "Energy secretary optimistic about Oil's Future," *New American*, July 12, 2020.

Associated Press. "Pariahs no more? U.S. reaches out to oil states as prices rise," *Greenwire*, March 10, 2022.

Cordesman, Anthony. "The myth or reality of US energy independence," *CSIS Report*, January 3, 2013.

Dillon, Jeremy. "Despite Trump's claim, U.S. still needs some Middle East oil," *E&E News*. January 10, 2020a.

Dillon, Jeremy. "Trump: 'We do not need Middle East oil,'" *E&E News*, January 8, 2020b.

Lee, Mike. "'A test of governments.' Oil industry faces worst glut ever," *E&E News*, March 18, 2020.

Lee, Mike, Carlos Anchondo, and Lesley Clark. "'We're going to have layoffs.' Oil industry braces for pain," *E&E News*, March 16, 2020.

Rapier, Robert. "U.S. net petroleum imports plunging toward zero," *Forbes*, March 21, 2018.

Richards, Heather. "Is the world headed to negative oil prices?" *E&E News*, April 2, 2020.

Tobben, Sheela. "Oil for less than nothing? Here's how that happened," *Bloomberg News*, April 20, 2020.

U.S. Department of Energy, Energy Information Administration (EIA). "U.S. crude oil production grew 17% in 2018, surpassing the previous record in 1970," April 9, 2019.

U.S. Department of Energy, Energy Information Administration (EIA). "Oil and Petroleum Products Explained: OIL IMPORTS AND EXPORTS," March 2021.

Yergin, Daniel. *The Quest: Energy, Security, and the Remaking of the Modern World*. New York: Penguin, 2011.

Yergin, Daniel. *The New Map: Energy, Climate, and the Clash of Nations*. New York: Penguin, 2020.

Q4. DID U.S. PLANS OVER SEVERAL ADMINISTRATIONS TO ACHIEVE "ENERGY INDEPENDENCE" BRING ABOUT A LARGE INCREASE OF SHALE OIL AND GAS SUPPLIES DURING THE 2010s?

Answer: Both domestic policy and global supply and demand played roles in the oil and gas shale revolution in the United States.

The Facts: This discussion briefly reviews oil prices after the turn of the twenty-first century, and then considers the pattern of domestic energy policy in the executive and legislative branches. It includes appointments in the new Trump administration and changes respecting federal lands and environmental regulation.

Oil price review. In capitalist societies, prices are a "trigger" for both demand and supply. From a low point of $20/barrel in the early 2000s, oil prices in the United States rose quickly in response to record levels of energy consumption and growth in oil imports. The price environment was optimal for development and expansion of shale oil and gas plays in Texas, Oklahoma, Louisiana, and the Appalachian regions. By the summer of 2008, the price of oil reached its peak at $147/barrel, reflected in gasoline prices over $4/gallon.

Thereafter, supplies surged while prices dropped. The price remained above $100/barrel for more than five years. In 2013, however, prices fell from $110/barrel to $75/barrel—and then further declined in early 2014 to $50/barrel. Prices remained at this level until early 2020, when they crashed due to the arrival of COVID-19 and a related, precipitous decline in economic activity around the world. Both beginning and ending the production of petroleum is a time-consuming process. As the oil market collapsed and financing for production declined, shale oil and gas producers cut costs and production—but not in time to ward off an oil and gas glut. While most drillers were burning off (flaring) surplus gas, some drillers in Texas pumped excess gas down wells (Cunningham, 2019).

Shale oil and gas are different from conventional petroleum resources, where deep wells are drilled to reach an abundance of hydrocarbons. Instead, the shale rocks are fractured, releasing relatively quickly exhaustible amounts of crude oil and gas. Thereupon drillers move on to new wells, which requires another spurt of investment, unless producers return dividends to shareholders. Called the "fracker's dilemma" in Bethany McLean's *Saudi America* (McLean, 2018), the shale business is a troubled one. The head of a large natural gas firm labeled the shale industry an "unmitigated disaster for the buy/hold investor" (whose strategy is passive—to ignore day-to-day fluctuations and allow the investment to grow with longer-term market trends): "The fact is that every time they put the drill bit to the ground, they erode the billions of dollars of previous investments they have made" (Richards, 2019).

This paradox is explained by the approach to growth taken by HF companies as compared to conventional oil/gas exploration and production (E&P). In the first years of hydraulic fracturing, large volumes of gas or oil are produced and the return to investors is very good; but to increase production further requires drilling more wells. When prices are good and increasing, dividends to shareholders can be reduced to finance drilling more wells. However, when prices are low and falling, the choice is between more wells or dividends, and HF company owners tend to keep employees working and cancel dividends. No longer do they have access to capital; and unable to weather the business downturn, they go bankrupt. The size of firms engaged in conventional oil/gas exploration and production is larger, and they have been better able to outlast price volatility (Hipple & Sanzillo, 2019).

Pattern of domestic energy policy. In several respects, the pattern of federal policy shows signs of responsiveness to the boom/bust cycle of the market in oil and gas, but it follows a different rhythm, determined by presidential and congressional election cycles and their outcomes. Taking office in 2001 at a time of low oil prices, President George W. Bush was a strong advocate for U.S. energy independence. In 2005, with Bush's encouragement, the U.S. Congress (then under Republican control) passed the Energy Policy Act, which legislated new incentives for oil exploration and production. In 2007, Congress enacted the Energy Independence and Security Act (which emphasized biofuels) domestically. That year President Bush said the United States was "addicted to oil" (Council on Foreign Relations [CFR], 2017).

Democratic presidential administrations and Congresses have followed energy independence goals, but have historically paid greater attention than Republicans to environmental protection values. During the Obama administration's eight years (2009–2017), for example, it imposed an effective moratorium on offshore drilling (a moratorium reinforced by the negative public response to the *Deepwater Horizon* oil spill disaster in the Gulf of Mexico in 2010) and on oil drilling in Alaska's Arctic National Wildlife Refuge (ANWR).

During Obama's presidency, however, the so-called "Arab Spring" revolutions that swept the Middle East in the early 2010s (including Libya, which had a spiking effect on oil prices) reminded U.S. policymakers about increased risks of instability in Middle Eastern oil regimes. The Obama administration took the fourth major drawdown from the U.S. Strategic Petroleum Reserve when OPEC ministers did not agree to raise output to blunt price spikes. In his 2014 State of the Union address, President Obama indicated that his energy policy was "all of the above." By this he meant that oil and natural gas would continue to be major fuels (albeit with increased efficiency in production and consumption). But he also pledged to increase solar and wind energy and increase innovation in the domain of renewables. He commented on the need to address climate change and significantly reduce carbon-based pollution, which was widely interpreted as critical of coal production and consumption (Plumer, 2014).

At the end of Obama's second term, Republicans controlled Congress, and they proposed lifting the oil-export ban, which with one exception had been in effect for 40 years. The White House and Democrats agreed to the plan in exchange for an extension of the renewable energy tax credit and other tax break priorities (House et al., 2015). A symbolic container of light crude from the Texas Eagle Ford shale field was the first shipment to Europe under this act. In the last year of his term, Obama earned the praise of environmental groups by rejecting a proposal to transport more than 800,000 barrels of oil/day from Canada to Texas via a proposed Keystone XL pipeline (CFR, 2017). (Similarly, President Biden won plaudits from natural gas producers for pledging to export more gas to Europe in support of regional energy security [Soraghan & Anchondo, 2022]).

Trump administration. Shortly after his election to the presidency in 2016, Donald Trump promised a new era of "energy dominance," a term he perhaps coined to imply that there would be no obstacles to full-scale federal action to greatly increase production of conventional as well as tight oil/gas (type found in impermeable shale and limestone, extracted using HF), to make coal "king" again for electricity generation, and to expand nuclear and hydropower capacity. This was a much more aggressive agenda than that of previous Republican presidents, and was seen as a clear response to the interests of Big Oil, allied economic sectors (e.g., construction, transportation), and constituent groups such as older Caucasian voters, lower middle class, and less well-educated workers. The Trump administration subsequently appointed a wide assortment of pro-development individuals to important government posts, expanded fossil fuel development activity on federal lands, removed or weakened regulations protecting the environment against pollution of air and water, removed or weakened regulations designed to protect threatened and endangered species and ecosystems, and increased provision of new infrastructure for fossil fuel development and consumption (Boylan et al., 2020).

Appointments. Initial appointments to his cabinet reflected President Trump's values. Rex Tillerson, former head of ExxonMobil, became secretary of state, indicating the linkage of foreign and energy policy. Rick Perry, the former governor of the oil-rich state of Texas, became secretary of energy. Former attorney general of Oklahoma Scott Pruitt, well known for his opposition to environmental regulation, became administrator of the Environmental Protection Agency (EPA). Finally, former Montana congressman Ryan Zinke became secretary of the interior. The executive director of the Alaska Wilderness League characterized Zinke in a way probably widely shared by environmentalists: "From the moment Secretary Zinke paraded in on horseback he's presided over a reign of terror for our most cherished wild places" (Brugger, 2018). All these officials had left office by late 2019 (both Zinke and Pruitt under the cloud of scandal); they were replaced by more experienced bureaucrats who carried on the pro-development, anti-regulatory work of their predecessors.

Federal lands. The Trump administration sought to expand access to federal lands for oil and gas drilling, coal mining, and mining of strategic minerals (such as rare earth elements). Most remaining U.S. public lands are in the West, and since the Sagebrush Rebellion, elected federal and state officials in Western states have resisted "federal overreach." (The Sagebrush Rebellion was a political movement in the 1970s and early 1980s in the 13 Western states in which the federal government owned between 20 and 80 percent of the land. The objective of Sagebrush leaders was to strip control of lands from federal agencies, such as the Bureau of Land Management.) Thus, the administration's efforts garnered praise from the oil and gas industry and related economic sectors, while environmentalists and a few other interests (cattle ranchers, fisheries) were in opposition due to fears that increased drilling and other operations might have negative impacts on their own livelihoods. Examples of these efforts to open up new public lands to energy exploration included expansion of offshore drilling in previously restricted areas of the Arctic, the California coast, and the Atlantic. ANWR has been the poster child of the U.S. environmental movement for nearly four decades, and the decision of the administration to allow oil/gas exploration in a limited zone (which the House and Senate had already agreed to) was met with anger and derision from drilling opponents. Also, lands once reserved from development out of concern for rare and threatened species (such as the western sage grouse) were opened as well, as were lands opened for mineral development (Marshall, 2021).

Removing or minimizing environmental regulations. Chapters 5 and 6 discuss in greater detail the Trump administration challenges to U.S.

environmental gains made in previous administrations. These challenges included attempts to limit or waive environmental regulations so that they did not impede energy and other apparently hazardous proposed developments, limits of the application of Endangered Species Act (ESA) protections of migratory routes, and limits on length of public comment periods for responding to proposed environmental impact statement (EIS) reviews. (Under terms of the National Environmental Policy Act of 1969, the federal government is required to evaluate positive and negative effects of proposed changes to the environment such as permitting construction of a road, a pipeline, or a dam, and list alternatives to the proposed action, before issuing a permit.)

New infrastructure. Another focus of the Trump administration was to invest in new infrastructure to move energy commodities to markets in the United States and abroad. An early action of the administration was to reverse President Obama's decision on the Keystone XL pipeline and sanction its construction and operation across state lines and the U.S.–Canada border (Soraghan, 2020). The administration also proposed construction of new ports and harbors to accommodate accelerated trade of liquified natural gas (LNG) to Asian countries. The Trump administration also recommended that railroads used to transport oil and coal be improved and new ones constructed (e.g., to link Alaska operations with Canadian lines).

Many of these actions fall under the rubric of "industrial policy" (discussed at length in Q9 of chapter 2), as they seek to make government the agent of change in a capitalist society, favoring some enterprises more than others because of the extent to which the policy is seen as benefitting national economic development. Other countries readily adopt industrial policies; for example, historically, Japan has favored large financial organizations and automobile firms to the exclusion of smaller ones. In the United States, industrial policy has been controversial because of the laissez-faire attitude toward government in the nineteenth century and the more limited role of the federal government in state affairs until the twentieth century. Nevertheless, both political parties have on occasion taken industrial policy positions. Virtually every member of the Congress has urged government to support the economic interests of their constituents, and presidents regularly issue executive orders and propose legislation advancing the interests of their election coalition.

Commenting on the question of which factors—"energy dominance," the market, technological change, or other elements—most determine outcomes, Laura Zachary, an economics consulting firm director, stated that most federal actions "have had marginal, if any, impact compared to larger market forces" as well as advances in technology. For oil, she said, price alone explained between 85 and 97 percent of the change in U.S. crude production during Trump's first year in office; "That's not success— that's just basic math" (Brugger, 2018).

Similar comments were made during the first year of the Biden administration (Richards, 2022). Prices of oil and gas rose to the highest level in nearly a decade. Profits of the majors rose as well, and producers increased production by 10 to 15 percent (Bloomberg, 2022; Lee, 2022). Progressive Democratic legislators accused large oil companies of profiteering and called for a windfall profits tax, but the American Petroleum Institute (API) said "political grandstanding . . . does nothing but discourage investment at a time when it's needed the most" (Lee & Anchondo, 2022).

In chapter 7, the Biden administration's actions in appointments and policy changes are considered.

FURTHER READING

Bloomberg. "Shale, seen adding a million barrels a day, is back to booming," *Energywire*, February 7, 2022.

Boylan, Brandon, Jerry McBeath, and Bo Wang. "U.S.-China relations: Nationalism, the trade war, and COVID-19," *Fudan Journal of Humanities and Social Sciences*, Vol. 14 (October 2020): 1–18.

Brugger, Kelsey. "Zinke's impact on 'energy dominance,'" *E&E News*, December 17, 2018.

Council on Foreign Relations (CFR). *Timeline: Oil Dependence and U.S. Foreign Policy*, 2017.

Cunningham, Nick. "US shale cautious as oil majors invade Texas," *Oil Price*, June 10, 2019.

Hipple, Kathy and Tom Sanzillo. "Risks to fracking companies in Appalachia Mount," Institute for Energy Economics and Financial Analysis, July 2019.

House, Billy, Kathleen Miller, and Erik Wasson. "Pelosi, White House support plan allowing U.S. crude oil exports," *Bloomberg News*, December 15, 2015.

Lee, Mike. "Exxon signals highest profit in 10 years," *Energywire*, April 5, 2022.

Lee, Mike and Carlos Anchondo. "Oil companies still see risks after billions in profits," *Energywire*, May 2, 2022.

Marshall, James. "Interior opens 10M acres to mining in northwest Alaska," *E&E News*, January 19, 2021.

McLean, Bethany. *Saudi America: The Truth about Fracking and How It's Changing the World*. New York: Penguin Random House, 2018.

Plumer, Brad. "Remarks of President Barack Obama—as prepared for delivery," *Washington Post*, January 28, 2014.

Richards, Heather. "Is U.S. shale facing an 'unmitigated disaster'?" *E&E News*, September 19, 2019.

Richards, Heather. "Drilling permits spiked then plunged under Biden," *Energywire*, March 14, 2022.

Soraghan, Mike. "How Trump's 'energy dominance' backfired on an $8B pipeline," *E&E News*, July 7, 2020.

Soraghan, Mike and Carlos Anchondo. "Biden's LNG deal with Europe jolts gas critics," *Energywire*, March 28, 2022.

Q5. HAVE MARKET CONDITIONS THAT FAVOR NATURAL GAS AND RENEWABLE ENERGY AND PLACE COAL AT A COMPETITIVE DISADVANTAGE BEEN TOO GREAT FOR PRO-COAL POLITICIANS AND THE COAL INDUSTRY TO EFFECTIVELY COUNTER?

Answer: Yes.

The Facts: The Trump administration compiled a consistent pro-coal record during its single term in office, whereas the Obama administration had a more mixed record. The Obama White House established policies designed to eventually wean the United States off coal, but took an incremental approach so as to limit the economic pain of such a move on coal-producing regions of the country.

In the end, though, neither administration's efforts had as great an impact on the coal industry as market forces, with utilities and other energy consumers turning away from coal due to its expense and adverse environmental consequences, ranging from polluted water to climate change. As a result, coal—which from 1950 to 2010 accounted for 50 percent or greater of the electricity produced in the United States (Kuckro, 2019)—appears destined to become a minor source of energy as America moves deeper into the twenty-first century.

Obama administration. During the 2016 presidential election campaign, candidate Trump criticized the Obama administration for its "war on coal." Trump pointed to the Obama administration's moratorium on coal leasing on federal lands. He also castigated regulations requiring expensive retrofits for coal power plants and the Clean Power Plan (a signature policy of Obama) designed to reduce greenhouse gas emissions from burning coal (discussed further in Qs 31 and 33). Trump claimed that when he was president, coal would once again become "king" of the U.S. energy economy.

In reality, the overall position of the Obama administration on coal and other fossil fuels was ambivalent. It emphasized renewables and supported carbon capture and other initiatives to mitigate climate change resulting from fossil fuel consumption. But the Obama White House also recognized that a sudden and dramatic turn away from fossil fuels posed significant economic and political risks.

During the Obama administration, a guiding thought was that coal was passé. The shale revolution (combining hydraulic fracturing with horizontal drilling to squeeze oil and natural gas from shale) sent prices of natural gas lower than coal, and major consumers of energy such as power plants were quick to notice. According to one 2018 estimate, natural gas was responsible for 92 percent of the decline in coal production and related job losses from 2008 to 2016 (Fell & Kaffine, 2018). But it was environmental factors in addition to price declines that triggered closures of more than half of the nation's coal mines. From a high point of 1,435 in 2008, they declined to 645 in 2017 (EIA, January 2019). Burning coal produces ash, which is hard to dispose of and leaches into rivers and lakes and contaminates groundwater. Burning coal also emits carbon dioxide and methane, two of the most dangerous greenhouse gases, which both the Kyoto Protocol and Paris Accords have sought to eliminate. Coal thus is the dirtiest of the fossil fuels, prompting the Sierra Club environmental organization to launch a Beyond Coal campaign with the specific aim of shuttering coal-fired power plants; and 2020 presidential candidate Michael Bloomberg promised to spend $500 million to close coal facilities and create a clean-energy economy (*E&E News*, 2019a).

Threats to coal. As use of coal has been attacked for both economic and environmental reasons, support for the fossil fuel has fractured and declined. The electricity-generating utilities, once firmly wedded to the coal industry, increasingly have argued that they can save consumers money by replacing coal with natural gas, producing cleaner energy that has fewer toxic emissions (Tomich, 2019). Once central to General Electric's (GE) balance sheet, coal accounts for a smaller and smaller share, and the company has stopped building coal power plants (Storrow, 2020). A number of the largest coal-fired power plants, such as several owned by Duke Energy, have closed over the course of the 2010s. Patriot Coal, which was once the largest producer of thermal coal in the Eastern United States, went bankrupt as well. States in the South that have historically used a large proportion of coal in their power plants also reduced their coal consumption, as did those in the Midwest and West. A number of the utility commissions across the country that have reduced their dependence on coal also made reference to using more wind and solar as growing parts of their electricity mix.

Trump administration. Once elected, President Trump faced formidable obstacles to making coal king again. He continued the combative stance used in the election campaign, and made abundant use of his Twitter account and executive orders to redress what he (and close advisors) saw as energy policy errors of the Obama and previous administrations. He listened to Bob Murray, founder and former CEO of Murray Energy, once the largest coal company in the United States who was a major donor to the Trump election campaign and a highly vocal critic of climate science and regulation of coal. The president held rallies in pro-coal regions of Pennsylvania and coal states such as West Virginia. Most significant, however, were appointments to head agencies and shape the agenda of regulatory change.

Appointments to energy. President Trump appointed Rick Perry as secretary of energy. This former two-term governor of Texas, a state with strong ties to the fossil fuel industry, made several attempts to revive the coal sector. He proposed federal intervention in electricity markets when owners of small, old, unprofitable coal plants sought to convert them into natural gas or renewable energy-powered plants. He also sought subsidies for coal from the Federal Energy Regulatory Commission (FERC). Champions of other fossil fuels (oil, natural gas) and renewables disagreed with these efforts, however, as did conservatives who opposed intervention in private-sector power markets, and administration officials who thought the national security arguments for protection of coal were weak (Northey & Behr, 2018). When Perry left the administration in November 2019, he was replaced by Dan Brouillette, an experienced federal bureaucrat, who said he would be emphasizing the power grid and its resilience. Asked whether any energy secretary could affect coal's future, Mary Anne Sullivan, a former general counsel of the department in 1998–2001, said: "I think the answer is probably not. I think Perry made the best effort that could be made with the very limited authority the department has" (Clark, 2019a).

Interior. President Trump's appointment to the Department of the Interior was Ryan Zinke, former Montana congressman and outspoken advocate of Western energy development. The Interior Department administers a large portion of the public lands in the United States (about 640 million acres), most of which are in the West, where nearly half of the land is federal domain. Interior's Bureau of Land Management (BLM) operates the federal coal-leasing program on public lands. Shortly after his appointment, Zinke cancelled the moratorium on coal leasing authorized at the end of the Obama administration and sought expedited review of an expanded leasing application (involving coordinated efforts of BLM and the Office of Surface Mining Reclamation and Enforcement [OSMIRE]). This action won plaudits from industry and business leaders in Wyoming and Montana and the National Mining Association, but objections from environmental organizations, several of which sued in federal district court to stop Zinke's bid. Judge Brian Morris ruled against Secretary Zinke and required Interior to prepare an environmental analysis of the proposed leasing changes (King, 2019). The secretary resigned after allegations of conflicts of interest, and was replaced by his deputy, David Bernhardt, who himself was questioned by former Obama administration officials and environmental nongovernmental organizations (NGOs) for his past years of lobbying on behalf of fossil fuel industry stakeholders (DiChristopher, 2019).

EPA. President Trump's appointment to the head of the EPA was Scott Pruitt, who had been a vocal advocate for rollback of federal environmental regulations (and particularly the Clean Air and Clean Water acts) as attorney general for the state of Oklahoma. Pruitt set the tone for deregulatory efforts of the Trump administration by criticizing agency programs, staff, and resources. For example, not finding what he regarded as "secure" communications systems in EPA offices, he authorized spending hundreds of thousands to build a secret calling room. Conflicts of interest and scandals ultimately forced his departure. He was succeeded by his deputy, Andrew Wheeler, who was more adept than Pruitt in winning concessions for coal mining and power plants. Under the Clean Air Act, EPA is required to set federal air standards for sulfur dioxide and nitrous oxide (which cause smog and respiratory difficulties). Wheeler affirmed the agency's overall reduction of these two pollutants' meeting standards, thus allowing several large-scale coal plants to increase emission of the pollutants (Reilly, 2019). But in fact, pollution reduction was the result of a decline in the number of coal-fired plants operating in the United States. Two EPA-proposed loopholes devised by Wheeler's administration increased waste from coal-fired power plants—the first by removing controls over release of toxic chemicals into waterways, and the second by delaying closure of storage ponds for coal ash (another contaminant of waterways and groundwater) (Beitsch, 2020; Stieb, 2020).

However, in the first days of the Biden administration, the new president issued a sweeping "nonexclusive" list of actions taken by the EPA and other agencies that would be reviewed. Called "Protecting Public Health and the Environment and Restoring Science to Tackle the Climate Crisis," the review included treatment of wastes from coal-fired power plants (Brugger, 2021).

FERC. The Federal Energy Regulatory Commission has up to five commissioners, appointed by the president. In the Trump administration, a majority were Republicans, and they predictably voted in support of fossil fuels. In late 2019, FERC adopted rules for the largest electricity market in the United States—13 states in the mid-Atlantic and parts of the Midwest—that would favor natural gas and coal. A Democrat on the commission complained that the rules were "a bailout" for fossil fuel companies that would "stunt the transition to clean energy resources" (Kuckro & Dillon, 2019).

The states' regulatory powers. State governments also have regulatory authority in the area of coal operations and environmental pollution. One example of legislation involving both regulatory scales is the Abandoned Mine Land Economic Revitalization (AMLER) program, the purpose of which is to "promote economic revitalization, diversification, and development in economically distressed mining communities through the reclamation and restoration of land and water resources adversely affected by coal mining" (HR 2156, 116th Congress). This pilot project has greater flexibility than much federal law and has gained support of both pro- and anti-coal mining groups (Brown, 2018). A major advantage of federalism is opportunities for state governments to experiment using programs like AMLER, and coal-reliant states such as Wyoming and New Mexico have done so (Storrow & Klump, 2019).

When Scott Pruitt, EPA administrator in the Trump administration, met with state energy officials, he declared that the federal government should have a "hands-off" policy toward closures of coal mines/plants and nuclear power plants. Instead of "weaponizing" the EPA, states should "take ownership of the regulatory process" (Swartz, 2019). One intent of this speech may have been to mobilize state officials to pressure federal agencies, including the FERC, to oppose coal plant closures (*E&E News*, 2019). Following Pruitt's advice would have put states in an adversarial position regarding renewables and NGOs supporting them, and this would have raised questions about the independence of public utility commissions.

Future prospects for coal. Moody's Investors Services reports that coal will occupy just 11 percent of U.S. power generation by 2030, as compared to the 2019 rate of about one-quarter of all power (Marsh, 2019). This sharp decline is primarily due to market factors, not state or federal regulation. (In summer of 2021, coal's share of U.S. power generation rose to 26 percent from 22 percent the previous year, and correspondingly, natural gas's share fell because of this commodity's higher price [Iaconangelo, 2021].)

Globally, on the other hand, prospects for coal look less dire. The International Energy Administration predicts that coal use will remain relatively stable until 2024 (IEA, 2019). This difference between projections for the domestic and international markets is important, and involves two dimensions. First, there is little prospect for marketing U.S. "thermal" coal (used to provide heat solely) abroad and limited domestic demand for it; but there exists a stronger domestic and potential foreign market for "coking" coal (aka metallurgical coal, used in the steel-making process). Second, however, to secure increased larger global market shares for metallurgical coal, financing for further infrastructure development will be needed. The United States lacks adequate port and harbor facilities for shipping coal (as well as liquefied natural gas [LNG]) to Asia. Reluctance of Asian banks to finance coal development projects means that most of this investment will need to come from home (Gronewold, 2019).

Energetic supporters of continued robust coal use make two arguments. They clothe coal with the "transition" mantle, saying that until battery storage capacity in green energy options improves significantly, solar and wind renewables will be unable to replace coal and nuclear power. The huge U.S. coal reserves are an essential cushion. Coal supporters such as Dan Brouillette, Trump's secretary of energy, also argued that there were non-heat uses of coal that should be explored further. "We can make carbon from it, we can extract rare earth metals from it," claimed Brouillette. "We can look at the residue . . . and pull critical materials for battery storage" (Clark, 2019a).

These arguments have failed to arrest coal's dramatic fall in popularity. President Trump, though, did not suffer politically from his inability to resuscitate the sagging coal industry. He maintained higher levels of approval in coal-dependent, politically conservative communities in West Virginia, Wyoming, Montana, and elsewhere than in many other parts of the country (Brugger, 2019). And coal was largely absent from Trump's 2020 reelection campaign.

Coal and COVID-19. One of energy scholar Vaclav Smil's summary observations applies to both coal and "energy transitions": "Remember that energy transitions are inherently prolonged affairs lasting decades, not years" (Smil, 2014). However, the COVID-19 pandemic may have affected the transition. Shortly after the coronavirus erupted, the National Mining Association's (NMA) CEO Rich Nolan wrote President Trump, Speaker of the House Nancy Pelosi, and Senate Majority Leader McConnell, asking for extensive federal assistance: "In a perilous time, the essential work of our coal miners to produce the fuel to keep the lights on and homes warm and the certainty and security provided by coal power is just what we need to keep the country moving forward." The NMA was not reticent. It asked under terms of the Defense Production Act that all of the coal supply chain—the mines, railroads, barges, and power plants—be designated "critical infrastructure," allowing their essential workers to avoid state/local lockdowns (Brown, 2020a).

The congressional coal caucus led by Rep. David McKinley (R-W.Va) repeated these demands and asked additionally that the Congress reduce the black lung excise tax (paying medical benefits to miners whose companies had gone bankrupt) and also the Abandoned Mine Land fee (paid to clean up abandoned mines). As one would expect, Democrats in the House opposed these additions, noting that they were "completely unrelated to the current crisis" and would reverse efforts to improve public health (Brown, 2020b). Government statistics showed coal falling even further behind renewables in its contribution to the energy market during the COVID pandemic, when overall demand for electricity dropped (Marshall, 2020).

President Trump did direct that the coal industry remain open during the pandemic, and it was added to the list of 16 industries deemed vital to national security. But this move was criticized by opponents who argued that coal was not "essential" for electricity production (given its decline from one-half of electric power generated to less than one-fourth). Within a month, the last proposed U.S. coal plant was halted by Georgia regulators, to the relief of the Sierra Club, which said that the ruling "represents the end of a bad idea." The group's senior representative of the Beyond Coal campaign, Stephen Stetson, added that he did not expect any effort to be made to revive the Georgia project, as there was "barely a market" for existing coal plants (Anchondo, 2020).

The Biden administration EPA required electric companies to inform it if they would make sufficient upgrades to meet federal standards or close their coal units (against the background plan to decarbonize the power sector by 2035). The response of the Southern Co. was that it would shutter about 55 percent of its coal fleet by 2030, as the company shifted to a net-zero electricity mix (Swartz, 2021; Swartz, 2022).

FURTHER READING

Anchondo, Carlos. "Regulators kill last proposed U.S. coal plant," *E&E News*, April 15, 2020.

Beitsch, Rebecca. "EPA rule extends life of toxic coal ash ponds," *The Hill*, July 30, 2020.

Brown, Dylan. "Congress spent millions to revive coal country. Did it work?" *E&E News*, November 15, 2018.

Brown, Dylan. "Coal industry asks Trump, Congress for coronavirus bailout," *E&E News*, March 20, 2020a.

Brown, Dylan. "Congressional Coal Caucus pushes for industry bailout," *E&E News*, March 20, 2020b.

Brugger, Kelsey. "Trump hasn't saved coal in W. Va. They don't care," *E&E News*, August 19, 2019.

Brugger, Kelsey. "Biden orders sweeping review of Trump regulations," *E&E News*, January 20, 2021.

Clark, Lesley. "Brouillette: 'There's a bright future for coal,'" *E&E News*, December 4, 2019a.

Clark, Lesley. "Trump hasn't saved coal. Can DOE?" *E&E News*, November 25, 2019b.

DiChristopher, Tom. "Trump's new interior secretary David Bernhardt, confirmed days ago, is now under investigation," *CNBC*, April 22, 2019.

E&E News. "Bloomberg announced $500M effort to shut coal plants," June 7, 2019a.

E&E News. "Industry pushed state regulators to lobby for power rescue," October 15, 2019b.

Fell, Harrison and Daniel Kaffine. "The fall of coal: Joint impacts of fuel prices and renewables on generation and transmission," *American Economic Journal: Economic Policy*, Vol. 10 (2) (2018): 90–116.

Gronewold, Nathaniel. "Asian banks shun coal, strangling U.S. exports," *E&E News*, May 24, 2019.

Iaconangelo, David. "U.S. coal use to jump, despite renewable shift—report," *E&E News*, May 12, 2021.

International Energy Organization. "Coal 2019: Analysis with forecasts to 2024," December 2019.

King, Pamela. "Judge: Trump can't OK mining without environmental reviews," *E&E News*, April 22, 2019.

Kuckro, Rod. "Shale gas surge dethroned coal power," *E&E News*, December 9, 2019.

Kuckro, Rod and Jeremy Dillon. "FERC throws support to fossil fuels in largest power market," *E&E News*, December 20, 2019.

Marsh, Joanna. "U.S. railroads could face billions in losses as coal demand slumps," *Freight Waves*, September 5, 2019.

Marshall, James. "Renewables beat coal for first time in 130 years," *E&E News*, May 28, 2020.

Northey, Hannah and Peter Behr. "Trump and coal: 'The boss wants what the boss wants,'" *E&E News*, October 17, 2018.

Reilly, Sean. "EPA about-face lets emissions soar at some coal plants," *E&E News*, September 26, 2019.

Smil, Vaclav. *Energy: Myths and Realities*. Washington, DC: AEI Press, 2014.

Stieb, Matt. "Trump admin makes it easier for coal companies to pollute in America again," *New York Magazine*, September 1, 2020.

Storrow, Benjamin. "GE about-faces, will stop building coal power plants," *E&E News*, September 22, 2020.

Storrow, Benjamin and Edward Klump. "Trump hasn't saved coal. Can states?" *E&E News*, February 28, 2019.

Swartz, Kristi. "Pruitt: We didn't have a 'punitive or weaponized' EPA," *E&E News*, September 25, 2019.

Swartz, Kristi. "Nation's 3rd-largest utility to shut down half of coal fleet," *E&E News*, November 5, 2021.

Swartz, Kristi. "Georgia Power plans to double renewables, ditch all coal," *Energywire*, February 1, 2022.

Tomich, Jeffrey. "The next coal war? Industry versus its customers," *E&E News*, August 2, 2019.

U.S. Department of Energy, Energy Information Administration (EIA). "Annual energy outlook 2019."

Q6. ARE TOTAL FEDERAL AND STATE SUBSIDIES GREATER FOR FOSSIL FUELS THAN FOR THE MAJOR RENEWABLES?

Answer: A qualified yes: Fossil fuels have received more subsidies in most years.

The Facts: Complete data on individual state subsidies, incentives, and tax credits to the different energy producers are lacking; information is relatively complete only at the federal scale. Three kinds of evidence support the judgment that to the present, fossil fuels have benefited more than green energy options from government subsidies and other forms of assistance: the different purposes for government assistance to the energy sector; the extent of incentives such as subsidies and tax breaks crafted for different industries in the energy sector from 1970 to 2018, keeping in mind the market share of each source; and government research and development (R&D) spending for each source.

Purpose of government assistance. A classical liberal viewpoint is that government should not interfere in private-sector activity, as that distorts the market. Yet governments throughout the world either control energy production (true of communist, socialist, and most authoritarian governments) or provide assistance to specific private-sector producers (which is called industrial policy). Reasons vary. Historically, coal, oil, and natural gas were the first energy commodities produced for mass public consumption, and fossil fuels quickly became indispensable for industrial civilization and national defense. It was obvious to policymakers that national security justified the exploitation of nonrenewable resources (Yergin, 2009).

Although wind, solar, geothermal, and plant-based energy sources have been earth's gifts since time immemorial, governments subsidized them only after the onset of the environmental movement in the late 1960s and the broad-scale change in consciousness accompanying it. National security based on fossil fuels certainly has conflicted with environmental protection emphasizing green energy. An additional reason that lawmakers provide tax incentives is to benefit one large corporation, or its trade association, for the purposes of encouraging economic growth and/or to secure influential support for their political careers (Engler, 1961).

Federal incentives, 1970–2018. The Congressional Research Service (CRS) is one of the most authoritative nonpartisan sources on federal spending. It has found that the majority of federal tax credits or other incentives from the 1970s (when policymakers began to use the tax code to promote energy security) to the mid-2000s benefitted fossil fuels. For example, in the late 1970s, two tax preferences given to the oil and gas industry accounted for all revenue losses, and three-quarters of losses in the early 1980s. Then lower oil prices in the late 1980s and a change in policy (a "free market" approach to oil/gas and other energy sources adopted by the Reagan administration) reduced federal tax revenue from fossil fuel industries. By the 1990s, unconventional fuels (shale oil and gas) gained most from tax credits. In the 2000s, preferential treatment of renewable energy production made up a larger portion of revenue losses. In the 2010s, more revenue losses have been associated with incentives designed to promote alternate fuels and biofuels (Sherlock, 2011).

However, the analysis does not tell us *why* various energy sources may receive different levels of federal financial support. Perhaps renewables appeared to receive more credits because they were lumped together with all domestic factories receiving deductions for production activities, or because the tax expenditure data do not specify energy type—for example, wind and not natural gas. Finally, the tax code does not spell out economic, social, or other policy objectives; invariably it is a political compromise influenced by the negotiating skills of numerous legislators and lobbyists. At best, one can make side-by-side comparisons providing descriptions only. For instance, in tax year 2017, fossil fuels produced were 77.5 percent of total energy produced in the United States, but accounted for just 25.8 percent of tax incentives handed out to the energy industry. In contrast, renewables (excluding nuclear) constituted 12.8 percent of 2017 primary energy production, but gained 65.2 percent of tax incentives. (Nuclear was 9.5 percent of the total but received just 1.7 percent of the tax incentives.) (CRS, 2019).

Federal research and development (R&D) expenditures. Information available on federal grants does provide a cumulative history. Most relevant are the records of monies allocated to energy technology in the Department of Energy (which was established by the Carter administration in 1977). From FY 1978 to FY 2018, a sum of $158 billion was spent, with those funds allocated as follows: renewable energy (17.5 percent), energy efficiency (15.6 percent), fossil energy (23.9 percent), nuclear energy (36.6 percent), and electric systems (6.4 percent). (Most U.S. coal and hydroelectric generating facilities were installed before 1990 and are covered only partially in these data.) The data do not include smaller amounts of R&D funding supplied by the Department of Defense and other agencies (Clark, 2018). They do show fossil fuels gaining marginally more support than renewables.

Discussion and debate about government assistance to the energy sector likely will continue, and advocates doubtless will continue to draw correlations between the price of energy commodities they prefer and different tax breaks and grants. In early 2020, the oil and gas industry suffered from historically low prices because of cutthroat competition between Saudi Arabia and Russia (both cut the price of their oil in an effort to increase their global share of the market), and this was compounded by the COVID-19 crisis. President Trump said the oil and gas industry was "one of the top of the list" for government assistance, and his administration prepared lines of credit, especially for the small and medium firms (Clark, 2020).

FURTHER READING

Clark, Corrie. "Renewable energy R&D funding history: A comparison with funding for nuclear energy, fossil energy, efficiency, and electric systems R&D," Congressional Research Service, June 18, 2018.

Clark, Lesley. "Details emerge on Trump's oil rescue plan," *E&E News*, April 30, 2020.

Engler, Robert. *The Politics of Oil: A Study of Private Power & Democratic Directions*. Chicago: University of Chicago Press, 1961.

Sherlock, Molly. "Energy tax policy: Historical perspectives on and current status of energy tax expenditures," Congressional Research Service, May 2, 2011.

Sherlock, Molly. "The value of energy tax incentives for different types of energy resources." Congressional Research Service (CRS), updated March 19, 2019.

St. John, Jeff. "The real deal on US subsidies," Wood Mackenzie Business, August 3, 2012.

Yergin, Daniel. *The Prize: The Epic Quest for Oil, Money & Power*. New York: Free Press, 2009.

Q7. ARE ELECTRIC VEHICLES (EVs) LIKELY TO REPLACE UP TO 30 PERCENT OF THE CURRENT MARKET SHARE OF GAS-GUZZLING TRANSPORT IN THE UNITED STATES BY 2045?

Answer: No. Although experts agree that it is feasible for electric vehicles to replace at least 30 percent of the American gas-driven fleet by 2045, that is unlikely to become a reality for a variety of economic, cultural, and political reasons.

The Facts: In late 2021, EVs had about 4 percent of market share of automobiles and trucks in the United States, and prospects are fair to good that by 2045 manufacturers will be able to produce enough hybrids and total electric cars and trucks to capture nearly one-third of the market. However, buyers are likely to be reluctant to purchase an EV until its benefits and costs are comparable to those of a car with an internal combustion engine. The electric vehicle transition question is complex, and requires discussion of several points.

Support for EVs. The U.S. auto industry appears to endorse this energy transition and is participating in it. General Motors (GM) launched the Volt (a hybrid electric car) years before other majors and followed it with the Bolt in 2018; it planned to add 20 all-electric brands by 2023. Ford's Focus model is a hybrid as well, and the company is developing an electric version of the F-150 (its most popular pickup) and had pledged to spend more than $11 billion to produce 40 full- or partially electric models by 2022. Fiat Chrysler (renamed Stellantis in 2021 after it merged with PSA Peugeot Citroen) committed to spend $10 billion on electric or hybrid cars by 2022, and Volkswagen indicated it would dedicate $50 billion toward EVs. Toyota Motor Corporation promised 10 new electric models by 2022. The most popular EV is Tesla's all-electric self-driving car, the model series X (Ferris, 2018).

Support for EVs was tested when in a cost-cutting action GM announced it would close five factories and lay off 14,000 workers in North America. In response, President Trump tweeted that he was considering ending EV subsidies (buyers received a federal tax credit of up to $7,500, applied to the first 200,000 vehicles sold): "General Motors made a big China bet . . . when they built plants there . . . don't think that bet is going to pay off. I am here to protect America's Workers!" (Joselow, 2018). GM leaders reiterated the importance to the firm of an "all-electric future" and made no changes in its plans. In the end, Trump never followed through on his threat. Meanwhile, most Democratic presidential candidates called for all-new passenger cars to be zero emissions by 2030. Agents for large utilities were enthusiastic at the prospect of having 20 to 30 percent of the auto fleet composed of plug-in hybrids or battery EVs. This would require significant capacity upgrades to the grid, related to the need for new infrastructure, a major source of utility profits (Ferris, 2020a).

Opponents are divided. Big Oil (huge integrated companies with operations in all phases of the industry, from exploration to refining) has the most to lose from a transition to electric vehicles in the long term, because about half of its production goes into gasoline, diesel, and other fluids powering the U.S. transportation sector. Yet multinational energy corporations such as ExxonMobil, Chevron/Texaco, Shell, and BP increasingly are diversified and invested in non-fossil fuels such as wind and solar. However, independent companies such as Hilcorp, Murray, and Apache are less diversified and more wedded to oil and natural gas, and they have thus shown greater opposition to EVs. Economic conservatives such as Tom Pyle of American Energy Alliance (AEA), meanwhile, oppose EV subsidies for artificially distorting the market and making U.S. taxpayers pay the bill. An assortment of other industries and economic sectors (such as proprietors of gas stations and convenience stores with gas pumps) have expressed concern about the potential impact of electric vehicles on their fortunes as well (Iaconangelo, 2019c).

Availability of critical metals. The International Energy Agency reports that increased popularity of EVs would reduce the supply of critical metals used in making them—especially cobalt and lithium—as demand dried up existing supplies (IEA, 2018). Research Director Indra Overland of the Center for Energy Research in the Norwegian Institute of International Affairs countered, however, that increased demand would stimulate exploration and production of these same metals. The U.S. Senate Energy and Natural Resources Committee has addressed the availability issue. Simon Moores, a battery analyst, testified that the United States was behind in the "global battery arms race," because China had constructed 70 mega-factories as compared to only a dozen in America. Lawmakers proposed an "American Mineral Security Act" to ensure availability in the future (Iaconangelo, 2019b). Parts of this proposed bill were included in the American Energy Innovation Act, itself contained within the omnibus relief bill passed in the lame duck session of the 116th Congress (Postelwait, 2020) (see also Q12). An additional issue is disruption in supplies of key metals needed to produce EV batteries, a vulnerability to supply chains exposed in the coronavirus pandemic (Willson, 2020a).

Technological change. A requirement of the EV transition is adequate storage capacity. Bloomberg New Energy Finance (BNEF) notes that power stored in lithium-ion batteries may soon be cheaper than power generated in plants powered by natural gas. The cost of such batteries dropped 35 percent per megawatt hour in the previous year (BNEF, 2019). A second analysis from the McKinsey Center for Future Mobility estimated that if cycled carefully, old lithium-ion batteries could retain more than three-quarters of their capacity; also, they could compete with natural gas suppliers such as gas-powered turbines, further promoting employment of renewable energy sources (McKinsey, 2019). The firm finds, however, that recycling batteries is dependent on the number of "end-of-life" batteries available, and this will always trail demand for EVs, expected to reach 23 percent of passenger car sales by 2030 (from 7 percent in 2020; Wells, 2021).

Research (from the Lawrence Berkeley National Laboratory and Carnegie Mellon University) indicated a potential new technology that would increase the life of EV batteries by five times and expand the driving range from 30 to 50 percent (Willson, 2020b). Researchers in China used black phosphorus combined with graphite and a polymer gel in a new lithium-ion EV battery. It took less than 10 minutes to achieve a maximum charge (lasting more than 300 miles), but whether a phosphorus-coated battery could be used practically requires further testing (Willson, 2020c).

Technological changes in the area of alternatives to lithium-ion have been rapid as well in 2020 and 2021. Bill Gates and Volkswagen (using their own resources and ARPA-E seed money) assisted an EV battery startup named QuantumScape. It claimed a major breakthrough, a "solid-state" battery (using solid materials instead of flammable liquids to enable charging and discharging). An alternative to lithium-ion batteries, researchers called the prototype a building block for batteries, which would be capable of recharging 80 percent in only 15 minutes. Significant testing remained before the prototype could be produced at industrial scale (Iaconangelo, 2020b). Then a startup supported by the Department of Energy designed a battery that was inexpensive and sufficiently durable to support the power grid by itself. Form Energy described its pilot project as an "iron-air" battery, which the company stated could store and deliver electricity for 100 hours, much longer than the lithium-ion batteries at one-tenth the price. Form Energy is partnering with a large steel producer, tapping into an existing supply chain (Iaconangelo, 2021a).

General Motors has backed a startup called SES Holdings, which has developed a lithium-metal battery large enough to power a car. The leader of the energy storage group at Oak Ridge National Laboratory (in Tennessee), Jagjit Nanda said the SES product would be a "bridge between current state-of-the-art lithium-ion, and solid-state battery tech." If successfully commercialized, such a battery would be a technological advance by producing lighter, more energy-dense cells, safer and able to cover an increased distance (Coppola, 2021). Finally, another alternative to the lithium-ion batteries moved closer to commercialization, as a breakthrough was reported in development of zinc-air batteries. Zinc is less costly to mine than lithium, less dangerous environmentally, and because the zinc battery does not catch on fire, it would be very useful for utility-scale energy storage (Willson, 2021).

A related area of technological change is charging stations, which critics believe are insufficient in terms of both number and recharging efficiency. Drivers seem unlikely to switch from gas to electric vehicles when gas stations are so abundant and charging stations rare. Also, most chargers are too slow. Current models far exceed the 10 minutes on average needed to gas up a car, and investors in charging services in the United States are not numerous (*E&E News*, 2019). Recently, GM pledged to fund construction of nearly 3,000 EV fast-charging stations in partnership with EVgo, one of the largest American charging networks (Ferris, 2020b).

A large difficulty in popularizing EVs is overcoming the imagery of the gas station—so familiar to generations of Americans using internal combustion engines and fueling them at a pump. For example, prices of gasoline are pegged to benchmarks like West Texas crude, but electricity has thousands of prices. Second, gas flows rapidly through the pump hose, but electricity trickles out the charger (at a power level of 50 kW, providing 3 miles of range/minute compared with 254 miles of range/minute for the average sedan). Third, gas stations have standardized plugs, accessible to all drivers, which remains an aspiration for EVs. Other problems include resistance to payment by credit cards, inconvenient operating hours and locations, and so on (Ferris, 2021b).

Ford took more cautious steps in electrification than GM, electrifying fewer types of vehicles, buying batteries from others, and investing less. This changed in 2021 when it announced it would build two new factories in the Southeast to make millions of batteries and electric pickup trucks a year. Its expenditure of $7 billion is the largest investment in the company's history. (It also committed to a training program for workers to prepare them to work on EVs.) The company's mass production of batteries will enable it to reduce per-unit costs to $80/kilowatt hour (compared to current average EV battery costs of $150/kWh), making it competitive with conventional cars (Ferris, 2021a).

The last technological change concerns creation of a national network for EV-charging, formed by the National Electric Highway

Coalition (composed of 50 Edison Electric Institute [EEI] members, TVA, and the Midwest Energy Inc.). Its first actions would fill gaps in the charging infrastructure along the interstate highway system. With anticipated growth of EVs from 2 million in late 2020 to 20 million by 2030, public fast-charging stations would need to increase tenfold (from the base of 10,000 currently). The $1.2 trillion bipartisan infrastructure bill enacted in August 2021 has $7.5 billion in federal funding for charging networks. Some $5 billion could be used by states for EV chargers; $2.5 billion would support EV stations and hydrogen, propane, and natural gas infrastructure (Behr, 2021). However, gridlock in Congress, the Ukraine invasion, inflation, and bureaucratic confusion unsettled this plan (*E&E News*, 2022).

The legal and regulatory environment. At the federal level, the 2020 election of Democrats to the presidency (and vice presidency) and enough votes in the Senate to pass simple legislation improved the environment for the EV transition greatly. The Biden administration's agenda for clean energy and EVs is bold and ambitious, addressed in Q8 and 9. Regulatory authority over electricity is stronger at the state and local levels than the federal level, and several states focused on change, as indicated in the following examples from 2018 to 2022.

California is perhaps America's most progressive state regarding the EV transition. In 2018, former governor Jerry Brown pledged the state would achieve zero-carbon electricity by 2045. Regional transportation authorities in San Diego, Los Angeles, San Joaquin, and the Central Valley of Stockton have proposed to become all-electric by the 2030s. The statewide California Air Resources Board (CARB) has adopted a rule pledging zero emissions for transit buses by 2040. The state's regulators have extended their scope from passenger cars and buses to delivery vans, heavy trucks, and tractor trailers, which will create a market for zero-emissions vehicles (Iaconangelo, 2019d).

The task force of Massachusetts' Republican governor, Charlie Baker, recommended a regional transportation emissions program to limit emissions. In Michigan, state officials crafted a master plan for the location and number of EV fast-charging stations. In Houston, Texas, a consortium of local officials, energy executives, and academics proposed a target of making 30 percent of new vehicles electric by 2030 (in a city where half of greenhouse gas emissions came from the transportation sector). Texas is in the middle of the states regarding sales of EVs, and experts regard the goal as technically feasible (Klump, 2019). Finally, New York resembles California in having a very aggressive clean transportation plan, with new mandates for converting buses and fleet vehicles to electric models. In

January 2020, the state Department of Public Service put utilities in charge of identifying the best sites for building new chargers and upgrading nearby grid infrastructure (Iaconangelo, 2020a).

Prospects for EVs. In 2018, commentaries on the future of EVs were not sanguine. The Rapidan Energy Group predicted a wavering of demand for EVs, because of the slowing of the Chinese economy and failures of climate negotiations. Economists, engineers, and transportation policy analysts in the MIT Energy Initiative conducted a three-year study of the transportation sector's future. In their view, costs of EVs were too high, and the battery packs seemed unlikely to fall below the threshold of $100/kilowatt hour by 2030. Although the authors foresaw increased demand, they did not think that more than one-third of vehicles would be electric powered by 2050 (Temple, 2019). With the passage of time, prospects seemed better. One example was Elon Musk's celebration of Battery Day, when he orchestrated a demonstration for investors (honking their horns from inside their Teslas lined up neatly in parking lot rows). By making bigger batteries, in smaller factories, sourcing lithium, and adding silicon to it to reduce cost, Musk described the innovations that would make batteries cheaper and drop by $25,000 the price of electric cars (Ferris, 2020c).

The most recent survey by financial advisory firm KPMG found top executives of the auto industry confident that they would continue to profit as sales shifted to electric vehicles. More than 1,100 executives at car and truck manufacturing firms (and their direct suppliers) answered questions. A quarter of the respondents were Americans, and a quarter were Chinese. The first observation drawn from the survey was that it was the first time in a decade of asking the question that leaders expressed nearly unanimous agreement that the shift to EVs would be fast. Two-thirds of the American executives said they were either "somewhat or extremely confident" that the industry would be profitable in the next five years (as compared to today). As to the share of EVs in the market, U.S. executives had different views—from 20 to 80 percent by 2030. As to future changes, respondents saw a pared-back role for car dealerships as Tesla's model of selling cars online caught on. By 2030, most respondents said, the prices of EVs and gas cars would have similar sticker prices. Over three-fourths of the respondents thought consumers would not purchase EVs if they were unable to charge up to 80 percent of a battery within 30 minutes. KPMG's head of automotive research Gary Silberg concluded his analysis of survey results saying: "There's no doubt in my mind that King ICE [the internal combustion engine] is being dethroned" (Iaconangelo, 2021b).

An electric road trip. Reporters for *E&E News* took a two-month, 8,000-mile trip in an electric car, asking how EVs might transform American transportation. They found problems of infrastructure (e.g., lack of charging stations), but the EV performed in most respects like a gas-driven car. They interviewed auto workers, auto dealers, and other drivers about their experiences. Most thought that the spread of EVs was inevitable (Ferris, 2019).

The United States has a car culture quite different from the European, and love for their gas guzzlers has reduced interest in EVs. A new study in 2022 projected that 70 percent of consumers would prefer a gas or diesel engine in their next car to an EV (Bond, 2022). Lowering the costs of EVs and increasing their benefits will likely make them more popular in the coming decades. At the onset of the Biden administration, there was a bloom of support for EVs, along with consciousness of the need to address climate change. The war in Ukraine may have the effect of increasing the interest in EVs. Certainly, it made creation of a nationwide charging network easier, as indicated by Tesla's proposal to open part of its Supercharger network for use by other electric cards (Ferris, 2022). Yet uncertainty and limitations in the supply of raw materials and semiconductors used to make batteries remain a challenge for EVs generally (Iaconangelo et al., 2022).

FURTHER READING

Behr, Peter. "Major U.S. utilities plan coast-to-coast, EV-charging network," *E&E News*, December 7, 2021.

Bloomberg. "GM's battery bet shows off lithium-metal cell that tops rivals," *E&E News*, November 4, 2021.

Bond, Camille. "Most Americans want a gas car, despite EV surge—report," *Energywire*, February 11, 2022.

Chandler, David. "How to get more electric cars on the road," *MIT News*, January 21, 2021.

Coppola, Gabriella. "GM's battery bet shows off lithium-metal cell that tops rivals," *Bloomberg*, November 4, 2021.

E&E News. "The world still doesn't have enough places to plug in cars," February 15, 2019.

E&E News. "State of the Union: All the energy takeaways," *Energywire*, March 2, 2022.

Ferris, David. "7 takeaways from a wild year for EVs," *E&E News*, December 21, 2018.

Ferris, David. "What we learned on the electric road trip," *E&E News*, November 13, 2019.

Ferris, David. "GM, EVgo plan giant U.S. charging network," *E&E News*, July 31, 2020a.

Ferris, David. "General Motors claims battery breakthrough," *E&E News*, September 10, 2020b.

Ferris, David. "Tesla says battery breakthroughs to bring $25K electric car," *E&E News*, September 23, 2020c.

Ferris, David. "Ford unveils record $11B bet on sprawling EV factories," *E&E News*, September 28, 2021a.

Ferris, David. "Can EV chargers act like gas stations? That won't be easy," *E&E News*, December 22, 2021b.

Ferris, David. "Tesla: We'll open our charging network for federal cash," *Energywire*, February 24, 2022.

Iaconangelo, David. "Murkowski, Manchin target lithium battery 'Achilles' heel," *E&E News*, May 3, 2019a.

Iaconangelo, David. "EVs could make 80% of gas stations unprofitable," *E&E News*, July 12, 2019b.

Iaconangelo, David. "Calif. readies first U.S. plan for zero-emissions trucks," *E&E News*, December 9, 2019c.

Iaconangelo, David. "Analysts see 'highly uncertain' future for EVs," *E&E News*, December 20, 2019d.

Iaconangelo, David. "Crisis will spark 'electrification of transport,'" *E&E News*, May 19, 2020a.

Iaconangelo, David. "Bill Gates-backed startup claims EV battery breakthrough," *E&E News*, December 9, 2020b.

Iaconangelo, David. "DOE-backed startup claims battery breakthrough for clean grid," *E&E News*, July 23, 2021a.

Iaconangelo, David. "1,100 auto execs: Here's where EVs are headed," *E&E News*, November 30, 2021b.

Iaconangelo, David, Miranda Willson, and Mike Lee. "War shakes up market for EVs, batteries," *Energywire*, March 9, 2022.

International Energy Agency (IEA). "Global EV outlook 2018," May 2018.

Joselow, Maxine. "GM doesn't care what Trump tweets," *E&E News*, November 30, 2018.

Klump, Edward. "Texas oil capital makes big bet on EVs," *E&E News*, October 16, 2019.

McKenzie Center for Future Mobility. "Improving electric vehicles competence," May 3, 2019.

Moore, Martha. "Should utilities build charging stations for electric cars?" *The Pew*, September 11, 2017.

Postelwait, Jeff. "COVID-19 relief bill contains clean energy policy act," *T&D World*, December 23, 2020.

Temple, James. "Why the electric-car revolution may take a lot longer than expected," MIT *Technology Review*, November 19, 2019.

Wells, Ester. "Report warns of 'supply imbalance' for EV battery materials," *E&E News*, December 16, 2021.

Willson, Miranda. "EV supply chain could see years of shortages," *E&E News*, June 9, 2020a.

Willson, Miranda. "Technology breakthrough could increase EV range, battery life," *E&E News*, July 23, 2020b.

Willson, Miranda. "'Breakthrough in performance': Battery charges in 10 minutes," *E&E News*, October 14, 2020c.

Willson, Miranda. "Scientists unveil battery breakthrough for energy storage," *E&E News*, January 8, 2021.

Q8. DO SOME EXPERTS EXPECT RENEWABLES TO ACCOUNT FOR THE MAJORITY OF AMERICAN ENERGY CONSUMPTION BY 2045?

Answer: Yes, some experts do believe that renewables, including wind, solar, hydro, and biomass, will supply most of America's energy needs by the start of the next generation.

The Facts: *General constraints.* Commentators on energy policy tend to be very sensitive to demand. When economic development slows globally, demand for all forms of energy drops, with greater declines in higher-priced commodities (such as wind and solar). Global growth slowed in 2019–20, with a noticeable dip in China's economy, which drove international investments in renewables down 14 percent. Yet investments grew in start-ups of electric cars and lithium-ion batteries (but lithium shortages threatened EV growth); and the U.S. Energy Information Agency (EIA) stated that the United States consumed more energy from renewables than coal for the first time in mid-2019 (EIA, 2019). The authoritative energy historian Daniel Yergin saw change in the U.S. energy mix occurring slowly: "We have a $20 trillion economy that rests on an energy foundation that's been developed over a century; it doesn't go away overnight" (Behr & Marshall, 2019). Yergin's observation is supported by popular sentiment. The Pew Research Center found that while a majority of Americans agreed that the United States should become carbon-neutral by mid-century, about 7 in 10 thought there should be a mix of fossil fuels and renewables. A major shift to renewables would, in the view of most, increase costs and jeopardize economic growth (Bond, 2022a).

Technology and infrastructure challenges. Technological innovations needed to improve use of renewables include batteries (both expanded storage of the lithium-ion battery and alternative long-term storage devices), research and development on fuel cells and renewable hydrogen, and both sites for and improved speed of charging stations. Of particular concern is finding a reliable means to deliver power to residences, factories, commercial establishments, and the power grid—which the IEA has forecast will be supplied mostly by electricity in mid-century at times when the wind does not blow and the sun does not shine.

The existing fossil fuel infrastructure is mostly paid for, but renewables require new facilities, equipment, changes in design, access to, and monitoring of new fuel sources. Electrification and developing emerging technology for renewable energy would cost around $173 trillion, notes BloombergNEF (Wells, 2021). ICF Resources, a consulting firm, found that transmission upgrades to accommodate large volumes of wind and solar saved electricity consumers money (Iaconangelo, 2021a). One unanticipated result of the decarbonization effort was utilities (especially in the Southeast) that added large amounts of natural gas to their grids to support renewable projects (Swartz, 2021).

Opportunities for entrepreneurs—economic, social, and political. Environmental nongovernmental organizations (NGOs), auto makers, and local/state governments would benefit from a significant increase in the consumption of renewables. Most environmental NGOs uniformly oppose the U.S. fossil fuel economy, and seek opportunities to move toward a zero-carbon emissions economy. Two large NGOs, the Sierra Club and Greenpeace, use this goal to increase their social capital among members and the broader public. The multinational U.S.-based oil company ExxonMobil purchased wind and solar power in the Permian Basin, as one means to sanitize its image among climate change activists. The automaker Honda purchased wind and solar power to offset greenhouse gas emissions at its U.S. plants.

The Trump administration rejected clean energy goals and opposed extending subsidies to renewables, while ardently championing fossil fuels, throughout Trump's single term in office. The election of Joe Biden changed the terms of debate completely. During the 2020 campaign, candidate Biden had argued for a "100 percent clean energy plan." The stimulus proposal he presented to the Congress had a price tag of nearly

$2 trillion and included a number of green energy elements. Yet Wood Mackenzie, a major energy consulting firm, concluded that the United States probably would manage to decarbonize only 66 percent of its electricity before 2035 (the year the Biden plan promised to reach 100 percent clean energy) (Iaconangelo, 2021b). Energy leaders at CERAWeek by S&P Global pointed out short-term challenges of high inflation and supply chain problems, making it impossible that decarbonization efforts would reach 100 percent by 2045 (Willson & Lee, 2022).

Stances by state and city governments in favor of renewables win favor from climate change activists. Chief among the states in this regard is California, which had imposed a state moratorium on new offshore drilling and several policies benefitting renewables. The governor and leading state legislators of Illinois have proposed major increases in electric power generated by renewables. Other liberal-leaning states such as New York, Connecticut, Washington, and Hawaii have followed suit. The Republican governor of Massachusetts, Charlie Baker, formed a task force to increase renewables, and supported its positive recommendations. By 2019, nearly 80 counties and cities across the United States had adopted a completely zero-carbon electric power system, meaning "committed to, or achieved, 100% clean electricity." Yet conservative red states such as Oklahoma plan otherwise. The state secretary of energy commented, "We've been pretty good at putting renewables on the grid . . . without driving up artificial costs by making us remove natural gas" (Klump, 2021).

Some energy companies have announced ambitious goals to become carbon free by the start of the next generation. In 2020, Duke Energy, which is the nation's second-largest electric utility, announced a goal of doubling its renewable energy capacity (wind, solar, and biomass) by 2025. The increased capacity would require regulatory updates, which the firm was optimistic about obtaining (Swartz, 2020). (See chapter 2 for more coverage of renewables and nuclear power in the twenty-first century.)

Impact of the COVID-19 pandemic. The coronavirus had adverse impacts on most clean energy projects. When President Trump withdrew the United States from the Paris climate agreement in 2017, mayors in more than 400 cities across America formed a climate coalition to reduce greenhouse gases (GHGs). They did so through development of a variety of projects, but since most of them were financed with local sales and income taxes, funding fell dramatically during the pandemic-induced economic downturn. When Biden assumed the presidency, his agenda strengthened air pollution rules quickly but moved more slowly on regulation of power plant air toxics (Brugger, 2021).

The energy sector as a whole sustained large losses in employment at the onset of the COVID-19 pandemic and associated economic slowdown. Data to mid-2020 indicated losses in both clean energy and fossil fuel jobs. In clean energy, a disproportionate number of job losses were in energy efficiency specializations because many office buildings closed, with employees working from home, and construction projects stopped or were delayed. In fossil fuels, greater job losses were sustained in the oil field (because of price crashes) and coal (shuttering of mines) sectors, while natural gas jobs were less volatile (Iaconangelo, 2020). In September 2021, there were about 62,000 fewer jobs in the oil field sector than before the pandemic, but some service companies were transitioning to construction and maintenance of renewable power projects (Lee, 2022). Q12 focuses on job changes (including partial recovery) in wind and solar.

Overall, carbon dioxide accumulation fell more than expected, for multiple factors—replacing coal with natural gas and increased electricity supplied by renewables (hydropower, wind, solar) among others. A success story was that the Southwest Power Pool (managing 14 states in the central U.S. grid) attained 90 percent renewables (Bond, 2022b). Yet Ryan Wiser of the Lawrence Berkeley National Laboratory's Electricity Markets and Policy group opined that the "low hanging fruit has, in fact, been picked" (Behr, 2021). The path toward net-zero emissions will be rockier than renewables' advocates have imagined.

FURTHER READING

Behr, Peter. "100% clean power? U.S. may be halfway there," *E&E News*, April 14, 2021.

Behr, Peter and Christa Marshall. "Moniz: 'A 100% renewable system is not realistic,'" *E&E News*, February 6, 2019.

Bond, Camille. "Report: Most Americans don't back 100% renewables," *Energywire*, March 2, 2022a.

Bond, Camille. "Central U.S. grid reaches 90% renewables for first time," *Energywire*, March 31, 2022b.

Brugger, Kelsey. "Biden agenda advances air regs; mixed progress elsewhere," *E&E News*, December 13, 2021.

Cassedy, Edward S. and Peter Z. Grossman. *Introduction to Energy: Resources, Technology, and Society*, 3rd ed. Cambridge, UK: Cambridge University Press, 2017.

Cordesman, Andrew. "The myth or reality of US energy independence," *CSIS Report*, January 3, 2013.

Iaconangelo, David. "Clean energy job losses could be permanent," *E&E News*, July 13, 2020.

Iaconangelo, David. "Transmission costs 'killing' renewables—report," *E&E News*, September 10, 2021a.

Iaconangelo, David. "Biden's clean energy, net-zero goals 'not feasible'— report," *E&E News*, September 17, 2021b.

Klump, Edward. "Will Okla. disrupt Biden's 100% clean energy goal?" *E&E News*, May 6, 2021.

Lee, Mike. "Where the oil industry is headed in 2022," *E&E News*, January 11, 2022.

Swartz, Kristi. "Duke Energy to double renewables by 2025 for CO2 target," *E&E News*, April 28, 2020.

Swartz, Kristi. "Can U.S. phase out natural gas? Lessons from the Southeast," *E&E News*, December 8, 2021.

U.S. EIA. "Electric power monthly report," June 27, 2019.

Wells, Ester. "The cost of net-zero? $173 trillion," *E&E News*, July 22, 2021.

Willson, Miranda and Mike Lee. "'Son of Build Back Better.' Energy CEOs eye renewables' future," *Energywire*, March 11, 2022.

2

❖

Major Renewables

Historically, hydro- and nuclear power have been used to generate electricity in the United States. They supplied 27 percent of the energy used to produce heat and light in 2020 (EIA, 2021), and they have the potential to increase this percentage greatly. Both have problematical qualities, however, that reduce their usefulness.

Solar and wind, on the other hand, produce a smaller fraction of the electricity used in the United States—about 7.6 percent in 2018 growing to 10.7 percent in 2020. They are among a half dozen energy sources that are bona fide renewables (Usher, 2019), and unlike hydro and nuclear power, they have fewer disabling qualities. Among the leading assets of wind and solar are their lack of greenhouse gas emissions. In an era of concern about climate change, they are at the top of the list of "green" energy sources; also, they do not pollute the environment with toxic substances.

An additional characteristic of wind and solar is that when compared to both fossil fuels and alternate energy sources identified in chapter 3 (e.g., corn-based ethanol), they have fewer "externalities." This is a term used by political economists to describe the effect on third parties (often construed as the broader general public) of harms issuing from an exchange between buyers and sellers—a harm that is not figured into the transaction price. Thus, the dangerous smoke from a coal-burning factory is not remedied by the factory owner or by the consumer (e.g., by a higher price for the steel produced, with proceeds distributed to those suffering from the air pollution). Typically, in such a situation, the government regulates the exchange (or disallows it, or imposes a tax on producer or consumers (Eisner et al., 2000; Gosling & Eisner, 2013; Miller et al., 2012).

The last question in chapter 1 asked what the expectations of experts were about renewables—whether by the start of 2045, it was possible that anyone would foresee that renewables would provide a majority of the energy consumed. This question was easy to discuss because only one expert needed to be of this opinion. The first query in this chapter, Q9, is more difficult to answer because the bar is set higher: it really concerns whether renewables realistically could be implemented within nearly two generations (by 2070), and whether they could power all of America's houses, factories, offices, vehicles, and generators.

The next two questions in this chapter (Q10 and Q11) consider hydropower and nuclear power. The issue of feasibility continues: could the current contributions of hydro and nuclear to the U.S. energy mix be doubled? Related questions include: What obstacles are likely to interfere? What impact would an increase in this kind of renewable energy have on fossil fuels? How has the change in administration following the 2020 presidential election changed the fortunes of these two energy resources?

Few energy resources can be employed for human needs without transformation. Q12 turns to materials essential for trapping wind and solar energy, and then storing the energy for use during peak times. The materials used in photovoltaic panels and batteries are not as abundant as reservoirs of water and deposits of natural gas. They include cobalt, lithium, cadmium, and other critical metals including some rare earth elements (REEs). The inquiry echoes several questions asked in chapter 1 about their availability and sufficiency, as well as infrastructure needs.

The final two questions of the chapter (Q13 and Q14) focus on conflicts between supporters of exploration, development, and production of nonrenewable resources, meaning fossil fuels primarily, and those who support prioritizing investment in renewables. These conflicts are played out in public spaces and concern the vast federal land domain, about 27 percent of all U.S. lands (CRS, 2017). Q13 examines land-use conflicts over resource development such as hydraulic fracking, encompassing protests and action campaigns of anti-fracking groups.

Q14 has a different focus—the ballot box. At the federal scale in the United States, there are no national referenda on issues affecting the public. U.S. citizens just vote in national elections for federal officials— members of the House and Senate and the president/vice president. But 24 state constitutions especially provide for initiatives or referenda and even propositions to be presented to the voters. Q14 answers such questions as: Have some states voted on ballot propositions to limit fossil fuels? Promote renewables? With what outcomes?

FURTHER READING

Eisner, Marc A., Jeff Worsham, and Evan J. Ringquist. *Contemporary Regulatory Policy.* Boulder, CO: Lynne Rienner Publishers, 2000.

Gosling, James J. and Marc A. Eisner. *Economics, Politics, and American Public Policy,* 2nd ed. Armonk, NY: M. E. Sharpe, 2013.

Miller, Roger, Daniel Benjamin, and Douglass North. *The Economics of Public Issues,* 17th ed. New York: Pearson, 2012.

Muller, Richard. *Energy for Future Presidents: The Science behind the Headlines.* New York: W. W. Norton, 2013.

U.S. Congress, Congressional Research Service (CRS). "Federal land ownership: Overview and data," Washington, DC: 2017. https://crsreports .congress.gov/product/pdf/R/R42346

U.S. Department of Energy, Energy Information Administration (EIA). "Electricity explained," March 18, 2021.

Usher, Bruce. *Renewable Energy: A Primer for the Twenty-First Century.* New York: Columbia University Press, 2019.

Q9. IS IT REALISTIC FOR THE UNITED STATES TO DEVELOP AN ENERGY SYSTEM ENTIRELY BASED ON RENEWABLE POWER BY 2050?

Answer: No.

The Facts: Ernest Moniz, who served as energy secretary in the Obama administration from 2013 to 2017, has said that "the idea we're going to have by 2050 . . . a 100 percent renewable system is not realistic." He went on to say: "It doesn't violate the laws of physics to do it. But that doesn't mean it is politically or economically implementable, and I think that is the issue" (Behr & Marshall, 2019). The reasons start with costs (of resources and infrastructure) of creating such a system, as well as historical trends in America's industrial policy. The private sector's interest in investment is treated, followed by changes in the political landscape, at national, state, and local levels.

Costs of resources and infrastructure. The costs of any transformative energy system would be colossal, running into hundreds of billions or even trillions of dollars. They might be several times greater than the current value of the U.S. gross domestic product (about $22 trillion in the first half of 2021). Not only would the renewables—solar, wind, biomass, hydropower, and others—have to be affordable to consumers, they also would need to be stored in safe containers (especially for nuclear waste) able to last millennia. Conversion costs from coal and natural gas to renewables, such as for the electrification of buildings, would also need to be factored in (Willson, 2021). Included in the cost would be overhead expenses and also expenditures to change regulatory systems when needed. Worley, a global engineering firm, said "the processes themselves" would have to be changed (Behr, 2021a). One factor that heartens renewable energy advocates, though, is a decline in projected costs of renewable energy for the bulk power grid (Tomich, 2021; see also Q20).

If the United States were a communist or socialist nation, these costs could be absorbed by the government. However, because it is a capitalist nation, leaders are reluctant to include energy costs along with other spending (e.g., public education, health, and safety) that over time have become a defensible part of the commonweal. This leaves only two other methods for a full commitment to renewable energy: industrial policy (government decides which energy resource to support) or private-sector investment.

The record of industrial policy. What makes industrial policy difficult to defend in a capitalist society with a long democratic history (and without a statist tradition, as found in Japan and Germany) is that there are winners and losers; those holding the purse strings will never forget the losers funded by government. The case of Solyandra made it difficult for the Obama administration to attract additional funding for development of solar panels. An earlier (and even more expensive) failure occurred in the Carter administration. At a time of high oil prices, President Carter sought to aid growth of low-cost equivalent energy sources. He persuaded the Congress to charter a Synthetic Fuels Corporation, with an initial budget of $20 billion in 1980. Less than $1 billion of the authorized amount was spent, and little synfuel was produced. Then oil prices dropped, removing the incentive for the operation (Bayrer, 2011; Rodrik, 2004).

Critics of U.S. industrial policy often lose sight of the successful outcomes of government investment, though (Light, 2002). One such stellar public investment was in the Defense Advanced Research Projects Administration (DARPA), which used federal monies to connect computers, establishing the foundation for today's Internet. A second example was the development of semiconductor gallium arsenide. This material has become critical to wireless communications chips in cellphones and satellites, among other uses (Graham-Rowe, 2008).

The evaluation of U.S. industrial policy typically varies by the political stances and ideologies of those doing the evaluating. Critics of Carter and Obama administration industrial policies were more often than not Republicans, who preferred fossil fuels over renewables (Clark, 2021). As noted below, when the Biden administration assumed power in January 2021, it pushed for federal policies to support green energy growth (Iaconangelo, 2021). Thereupon, supporters of fossil fuels from the Trump administration quickly opposed planning (and proposed legislation) to boost renewables. However, the partisan divisions that dominate American politics today increasingly are cross-cut by age: older people are more opposed to phasing out fossil fuels entirely than younger people; younger Americans also are significantly more likely to support addressing climate change (and more likely to be activists on the issue and in social media) than older generations (Sobczyk, 2021).

Given these partisan differences, it came as a surprise to many observers in June 2021 that Republicans joined Democrats in the passage of "the most expansive industrial policy legislation" in U.S. history (Sanger et al., 2021). What motivated bipartisan cooperation was the competitive threat from China (particularly supply chain control in solar/wind technology, battery cells, and rare earth elements or REEs). The nearly quarter-trillion-dollar package included federal government investments in biomedical research, artificial intelligence, semiconductor manufacturing, and the like (see also Q12).

Private investment. The recent pattern of investments illustrates changing priorities respecting renewables. Private investment firms invested several billions in lithium-ion batteries in mid-2019, which then were sufficient to maintain solar/wind production temporarily. The prices of utility-scale batteries dipped, and several firms invested in them as a backup option to solar power. The four- to six-hour capacity of these batteries, consultant firm Wood Mackenzie observes, was sufficient to maintain solar/wind production for a time. A Swiss firm (Energy Vault) invested millions into commercialization of continuous baseload power, and domestic firms also directed capital toward enhanced connectivity (Iaconangelo, 2019). (The energy department's "super-grid" study of renewables established groundwork planning for country-spanning high-voltage, direct current power lines [see Q 20].) In addition, the bipartisan infrastructure law passed in 2021 had "tens of billions in new grants and loans" overseen by the Department of Energy, which likely will attract private-sector interest and support [Willson et al., 2022].)

No sign of the attractiveness of renewables is more meaningful than the changes evident in the investment dynamics of Big Oil. Royal Dutch Shell, a multinational giant of the oil and gas business, announced in April 2020 that it would make its complete business model carbon free by 2050 or earlier. CEO Ben van Beurden said: "(At) this time of extraordinary challenge (pandemic and oil price crash), we must also maintain the focus on the long term" (Gronewold, 2020). British Petroleum (BP) had been nearly two decades ahead of Shell when, in 2002, its CEO, John Browne, rebranded it as "beyond petroleum" and pledged to hold emissions constant. However, the expense of cleaning up major oil spills caused it to divest from its wind and solar assets until it pledged to return to renewable investing in 2020 (Carpenter, 2020). (A recent study suggests, however, that none of the majors' entry into renewables has been at a scale "that would indicate a shift away from fossil fuels" [Anchondo, 2022].)

ExxonMobil followed a different trajectory. It was the world's richest corporation in 2013 but lost that distinction by 2020. As other fossil fuel companies changed investment strategies when oil prices dropped precipitously in late 2019, ExxonMobil did not. A group of activist investors sought changes in the board of directors in response to climate change. Its letter said that Exxon's stock price and dividend rate had declined 20 percent in the previous 10 years, when the S&P index rose 277 percent. In late May 2021, the emergent investors won seats on the Exxon board, while Chevron shareholders forced its board of directors to broaden emissions reduction goals (Lee & Anchondo, 2021a). (Several months later, ExxonMobil announced it planned to reach net-zero emissions in the Permian Basin by the end of the decade [Lee & Anchondo, 2021b].) Meanwhile, shareholders of ConocoPhillips stock also voted to make the company's emissions reduction plans more ambitious.)

A final example is that of NextEra Energy., the world's largest U.S.-based utility company, which has had to manage problems in supply chains and keep construction sites open in the solar and wind industry during the pandemic. One of its subsidiaries, NextEra Energy Resources, signed multi-megawatt contracts in the three months after the onset of the pandemic (Swartz, 2020). By mid-year, it began planning a hydrogen pilot project, which firm leaders argued would further its goal to become a zero-emissions company. These are promising developments, particularly when viewed in combination with early steps of the Biden administration.

Biden administration changes. At the national level, the new Democratic administration announced industrial policies in early 2021 that were as strongly supportive of clean energy as the Trump administration had been in defending fossil fuels. President Biden said the United States would be "carbon free by 2035," an aspirational goal that the administration backed up with concrete steps such as appointments of climate change experts to important agency positions, reversals or freezing of concessions to the fossil fuel industry, and support for incentives to clean energy industries (including EV production).

Appointments as senior advisors included former Obama secretary of state John Kerry to post-Paris Accords climate agreements and former Obama EPA administrator Gina McCarthy to coordinate domestic climate change policy. Michael Regan, the head of the North Carolina Environmental Agency, was named EPA administrator, and a number of career civil servants moved into high positions at important agencies related to environmental and energy issues. However, diversity and economic justice goals of the administration led to several unexpected choices, such as Deb Haaland as Interior Department secretary. The first American Indian to be appointed to head a cabinet agency, Haaland had served in Congress representing New Mexico and been an activist in anti-fossil fuel demonstrations (Cho, 2021).

Reversals of fossil fuel actions occurred in those areas where the Biden administration could announce a new clean energy/green agenda without reliance on other branches of government. For example, within hours of his inauguration, President Biden ordered a review of regulatory rollbacks that the Trump administration had instituted to increase oil/gas production. He also froze Trump's opening of oil/gas leasing on BLM lands (although this decision was later reversed on legal grounds by a U.S. 9th Circuit Court).

Stimulating development of renewables took several forms. To decrease carbon dioxide emissions and reduce energy inefficiencies, the new administration vowed to make federal purchasing decisions to increase clean (EV) vehicles in the federal fleet and support clean buildings. The administration planned to increase manufacturing of solar and wind in the United States and with union labor (*E&E News*, 2021). It emphasized this when it announced in February 2022 a "Made in America" supply chain drawing on more than a dozen DOE offices and national labs. Observers such as Pol Lezcano, a BloombergNEF solar analyst, however, asked, "To what extent will policymakers take this seriously and actually try to implement these policies, and put actual dollars to work?" (Iaconangelo & Holzman, 2022). Because renewables require a larger footprint for collection/production than oil and gas, the Biden administration made space needs integral to U.S. infrastructure renewal. Its slogan "30x30" meant 30 percent of the American landscape was to be protected (from economic and especially industrial development) by 2030. This attracted both traditional preservationists and progressive environmental interests to the campaign (Behr & Tomich, 2021).

The Biden administration's fiscal year (FY) 2022 budget request included large increases to DOE's clean-energy research portfolio (including areas first included in Trump requests) as well as to EPA and the Interior Department. Discretionary spending to these three agencies would increase by 9 percent over the previous year under Biden's budget. Areas highlighted included nuclear, offshore wind, energy storage, EVs, and research and development (R&D) funding in a "whole of government" approach to taking action on climate change (Dillon, 2021). A significant part of the approach was the Clean Energy Standard (CES), which climate advisor Gina McCarthy claimed would decarbonize 80 percent of the power grid by 2030 if passed into law (Sobczyk & Waldman, 2021). The CES, however, did not survive opposition from fossil fuel lobbyists and Republican legislators, and Democrats eventually had to take it out of the legislation in order to get GOP support for the large infrastructure spending bill in which it had been placed.

State and local changes. Some local and state governments have championed energy policy reforms that emphasize investment in renewables and other steps to reduce carbon emissions. In South Carolina, the legislature enacted a Solar Energy Tax Credit, which remitted to taxpayers 25 percent of purchase and installation costs. But Maryland's Climate Solutions Act failed to pass the assembly after two years of effort. New Mexico, meanwhile, passed a 2019 Clean Electrification Act to steer the state away from fossil fuels (Swartz & Klump, 2021; Willson et al., 2021).

Yet both legislatively and through the ballot box, some states and local areas prohibited or strongly discouraged development of renewables. The Sabin Center for Climate Change Law of Columbia Law School reported that about 100 states, counties, and cities adopted policies and ordinances restricting development of wind and solar facilities. In Florida, the legislature enacted an Energy Preemption Bill, which erased hopes of residents in Tampa and other cities who favored renewables. Many of the restrictions were zoning limitations, such as setback requirements specifying distances of projects from other facilities or restrictions on where they could be situated. In some cases, objections came from fossil fuel lobbyists such as Americans for Prosperity, financed by the conservative Koch Brothers and other big contributors to GOP causes and politicians (Marsh et al., 2021).

California, America's most populous state, leads the nation in production of renewables, and it has the strongest regulatory and legislative basis of support for clean energy. By law, the state must meet carbon neutrality by 2045. A number of challenges exist to meeting that goal, such as weaknesses in permitting and construction of battery storage for solar and wind and other technologies. While stressing the difficulty of attaining clean energy goals, however, analysts also pointed out the reforms would "bring social benefits, such as less air pollution and improved public health. It will also create more jobs" (Elkind et al., 2020). The state of New York made a similar pledge to convert to zero-carbon electricity by 2040, but it remains an open question as to whether that is a realistic goal (Behr, 2021b).

Finally, a study from Ascend Analytics released in August 2021 describes the likely future scenario for Arizona, a sentinel state in the debate over clean energy standards both nationally and in the American states. The study described clean energy progress as "achievable and cost-effective" in the next two decades, but warned that the last 10 to 20 percent of decarbonization might be costlier and more challenging than the easier first steps (Klump, 2021).

The Biden budget proposals for fiscal 2023 were comprehensive and bold. The blueprint promised change of the energy sector to spur growth of wind and solar, modernize the grid, improve cybersecurity and pipeline safety—among other objectives not achieved in the "Build Back Better" act or infrastructure legislation (Richards et al., 2022). Definitely, this heavily loaded agenda faced difficulty in Congress (Anchondo et al., 2022) at a period of extreme partisan polarization and just before the 2022 midterm elections (when Republicans stood a good chance of flipping the House and Democrats of losing their razon-thin grip on the Senate).

Also, prospects for a clean energy future would seem to be highly dependent on technological innovation (such as cost declines in renewable energy technology [Wells, 2021]), in addition to economic assistance. Q10 starts discussion on major renewables by treating hydropower.

FURTHER READING

Anchondo, Carlos. "Are oil majors greenwashing? What 12 years of data show," *Energywire*, February 17, 2022.

Anchondo, Carlos, David Iaconangelo, and Mike Lee. "3 takeaways from Biden's energy plan," *Energywire*, April 1, 2022.

Bayrer, Ralph. *The Saga of the U.S. Synthetic Fuels Corporation: A Cautionary Tale.* Washington, DC: Academia/Vellum, 2011.

Behr, Peter. "Report details blistering pace of renewables needed for net-zero," *E&E News*, August 3, 2021a.

Behr, Peter. "N.Y. grid chief: 2040 carbon goal 'extremely challenging,'" *E&E News*, May 6, 2021b.

Behr, Peter and Christa Marshall. "Moniz: 'A 100% renewable system is not realistic,'" *E&E News*, February 6, 2019.

Behr, Peter and Jeffrey Tomich. "Biden's dilemma: Land for renewables," *E&E News*, March 24, 2021.

Carpenter, Scott. "After abandoned 'beyond petroleum' re-brand, BP's new renewables push has teeth," *Forbes*, August 4, 2020.

Cho, Renee. "A guide to the Biden administration's all-of-government approach to environmental justice." In *State of the Planet*, Columbia Climate School, March 4, 2021.

Clark, Lesley. "GOP support for renewables plunges," *E&E News*, June 14, 2021.

Dillon, Jeremy. "R&D funding, nuclear, renewables get big budget boosts," *E&E News*, May 28, 2021.

E&E News. "How Biden's order hit EVS, oil and clean energy," January 28, 2021.

Elkind, Ethan, Ted Lamm, and Katie Segal. "Capturing opportunity: Law and policy solutions to accelerate engineered carbon removal in California," Berkeley, CA: Center for Law, Energy & the Environment; Los Angeles: Emmett Institute on Climate Change & the Environment, December 2020.

Graham-Rowe, Duncan. "Fifty years of DARPA: Hits, misses and ones to watch," *New Scientist*, May 15, 2008.

Gronewold, Nathanial. "Shell plans to go carbon neutral by 2050," *E&E News*, April 18, 2020.

Hydro Review. "U.S. Senators introduce bill to incentivize hydropower upgrades, restore river flow," August 21, 2021. https://www.hydroreview.com/regulation-and-policy/u-s-senators-introduce-bill

Iaconangelo, David. "The conundrum: 100% renewables and energy storage," *E&E News*, October 24, 2019.

Iaconangelo, David. "Renewables break records, but where are the jobs?" *E&E News*, March 16, 2021.

Iaconangelo, David and Jael Holzman. "Biden admin releases blueprint for clean energy supply chain," *Energywire*, February 25, 2022.

Klump, Edward. "Ariz. study renews debate over 100% clean electricity," *E&E News*, August 23, 2021.

Lee, Mike and Carlos Anchondo. "What Exxon, Chevron climate shake-ups mean for oil," *E&E News*, May 27, 2021a.

Lee, Mike and Carlos Anchondo. "Exxon announces plans for net-zero emissions in Permian Basin," *E&E News*, December 7, 2021b.

Light, Paul. *Government's Greatest Achievements: From Civil Rights to Homeland Security*. Washington, DC: Brookings Institution, 2002.

Marsh, Kate, Neely McKee, and Maris Welch. *Opposition to Renewable Energy Facilities in the United States*. Sabin Center for Climate Change Law, Columbia Law School, New York, February 2021.

Richards, Heather, Miranda Willson, Carlos Anchondo, Christian Vasquez, and Kristi Swartz. "How Biden's budget would change the energy sector," *Energywire*, March 29, 2022.

Rodrik, Dani. "Industrial policy for the twenty-first century," KSG Faculty Research Working Paper Series RWP04-047, November 2004.

Sanger, David, Catie Edmondson, David McCabe, and Thomas Kaplan. "Senate poised to pass huge industrial policy bill to counter China," *New York Times*, June 7, 2021.

Sobczyk, Nick. "Poll: Fossil fuel phaseout unpopular across age, party divide," *E&E News*, May 26, 2021.

Sobczyk, Nick and Scott Waldman. "White House escalates push for clean energy standard," *E&E News*, June 30, 2021.

Swartz, Kristi. "World's largest renewable developer thrives as industry reels," *E&E News*, April 23, 2020.

Swartz, Kristi and Edward Klump. "How states may drive—or foil—Biden's clean energy plan," *E&E News*, February 5, 2021.

Tomich, Jeffrey. "Grid operator: 50% renewables challenging but 'achievable,'" *E&E News*, February 11, 2021.

Wells, Ester. "Report outlines path to 80% clean power," *E&E News*, September 9, 2021.

Willson, Miranda. "Phaseout of natural gas may spike utility bills—report," *E&E News*, August 2, 2021.

Willson, Miranda, Heather Richards, David Iaconangelo, Hannah Northey, and Edward Klump. "What to watch at DOE, FERC, Interior in 2022," *E&E News*, January 5, 2022.

Willson, Miranda, Kristi Swartz, and Edward Klump. "4 state trends remaking U.S. electricity," *E&E News*, July 9, 2021.

Q10. IS IT FEASIBLE TO DOUBLE THE AMOUNT OF HYDROPOWER IN THE U.S. ENERGY MIX IN THE MID- TO LONG TERM (25–50 YEARS)?

Answer: Yes, but only with considerable assistance from government and the private sector.

The Facts: Flowing water was used from the late nineteenth to the early twentieth century to turn paddle wheels, which powered flour mills and other small factories. Today, the power of water is harnessed on a much greater scale. By the third decade of the twenty-first century, hydropower plants almost exclusively produce electricity; running water spins turbines connected to generators. In 2022, 7.3 percent of electricity in the United States is produced by hydropower, and it contributes more electricity to the nation than any other of the renewables except wind (8.4 percent). The volume of hydropower-generated electricity, measured in gigawatts (GW), has increased, but the amount produced by other energy sources (and particularly natural gas) has risen even faster.

Status of hydropower production. By the end of 2015, the United States had 2,198 active power plants and 42 pumped storage hydropower (PSH—a type of hydroelectric energy storage) plants, for an overall capacity of 101 GW (EERE, 2015). If there were normal growth in this form of energy resource, by 2050 capacity would reach more than 200 GWs. The U.S. Bureau of Reclamation, which is in charge of many hydropower sites, reports: "Potential sites for all types of hydropower exist that would double the U.S. hydroelectric production *if they could be developed*" (emphasis added) (USBR, 2016). Some potential sites have "unfavorable terrain" and others are at the design stage. Both types of these sites would require expensive engineering work to develop.

Issues of ownership. Nearly three-fourths of the U.S. hydropower capacity is publicly operated, but ownership is diverse. Three federal agencies—the Army Corps of Engineers, Bureau of Reclamation, and the Tennessee Valley Authority (TVA)—own about half of the capacity. A number of other government bodies at different scales own a quarter of the nation's overall capacity, including states, public utility districts, irrigation districts, and rural cooperatives. Finally, slightly more than a quarter of capacity is in private hands, including utilities owned by investors, private companies, and private power producers. Given this diversity, it is sometimes hard to determine which body has the authority to make an essential upgrade. For example, the Bonneville Power Authority (BPA) technically is under the Energy Department's Bureau of Reclamation, but it operates almost independently, raising concerns about its accountability (Jacobs, 2019a).

Aging infrastructure. Many of America's large dams were built in the 1930s to address conditions of drought or flooding and also to employ people during the Great Depression. Congress authorized TVA in 1933, during the first 100 days of President Franklin D. Roosevelt's New Deal, and this led to construction of nine dams to control flooding on the Tennessee River. The BPA was built in 1937 to provide inexpensive power to the Pacific Northwest. In 2019, a cracked lock at Bonneville Dam closed down barging of wheat on the Columbia River, a major artery for farmers shipping grain abroad (Jacobs, 2019b). The BPA already has a debt of $15 billion, and to pay for repairs would require customers to pay higher rates—at a time when hydropower costs have become less competitive than those of natural gas and even coal. The elderly infrastructure of dams nationwide constrains their ability to attract investment for upgrading.

Adverse environmental effects. Dams halt the passage of water including all its contents, and the reservoirs formed behind dams cause water contents to settle, called *siltification.* Unless frequently dredged, riverbeds rise and then water flows over the dam. Formation of layers of silt also changes the riverine ecosystem, affecting fish, birds, and microorganisms. Dams also interrupt the migratory paths of fish and other aquatic species. Fish ladders can mitigate this interruption for some species, such as salmon, but not for all, and this remedy and other abatements (such as tagging fish and carting them downriver) are quite expensive (Reisser & Reisser, 2019).

A proposal to expand the U.S. hydropower system. In 2015–16, the DOE's Water Power Technologies Office (WPTO) assembled more than 300 specialists from hydropower industry firms, federal and state offices, environmental groups, universities, and other agencies. Their charge was to model hydropower's potential and to evaluate future means to develop low-carbon, renewable hydropower production (and also pumped storage) in the United States in the next generation (approximately 2025 to 2050). They were to take into account social and environmental factors and consider ways to reduce greenhouse gas emissions and other adverse environmental impacts.

The resulting "hydropower vision" estimated that U.S. hydropower could increase "to nearly 150 GW by 2050 (a growth of 50 percent), through upgrades to existing plants, adding power at existing dams and canals, and limited development of new steam-reaches (and) . . . new pumped storage capacity." The benefits would include billions of dollars in new economic investment, reduction of cumulative greenhouse emissions by 700 million tons carbon-dioxide equivalent, and some 75,000 hydropower-related jobs across the nation by 2050. The study forecast that about 35 million homes could be powered by hydropower via such investments (WPTO, 2016). Even if all this investment occurred, however, the total amount of hydropower generated would not even double from its current level. Nevertheless, the WPTO's latest report optimistically comments "U.S. hydropower capacity continues to grow through upgrades to existing plants and other types of innovative new projects" (EERE, 2021).

Sustainability issues. Congress has not kept pace with the deferred maintenance needs of large federal projects Certainly, very few megaprojects are on the drawing boards, because of economic and geographic constraints, environmental issues, and negative public perceptions about large projects.

The private-sector hydropower projects tended to be smaller, and for this reason have garnered support more easily (Bracmort et al., 2015).

A greater sustainability issue for U.S. hydropower concerns PSH (pumped storage hydropower), which contributes nearly 95 percent of energy storage resources. This system joins two water reservoirs at different elevations. Power is generated as water moves down through a turbine, and then the power is drawn to pump water (recharge it) to the upper reservoir. In 2021, residents and the local government of a community with a private PSH facility in the Catskill Mountains claimed it could lead to flooding and displacement of residents as well as threaten recreational and landscape values, causing the firm to withdraw its permit application (Willson, 2021). Declining costs of battery storage were a consideration as well.

The most recent threat to the sustainability of hydropower has been severe droughts, especially in the American West, largely attributable to climate change. The droughts have dried up rivers and reservoirs essential for production of zero-emissions hydropower. In August 2021, the California water project was forced to shut down a hydroelectric power plant at Lake Oroville because of low water levels. Water levels dropped 25 to 30 percent at other reservoirs such as at Lake Shasta and at the huge Hoover Dam on the Colorado River (Reuters, 2021). Lower water levels meant higher utility charges as water turbines have to pump harder and longer to meet consumers' needs. In California, electricity prices jumped 150 percent in three months' time. A disappointment to observers of the transition to clean energy is that some of the largest beneficiaries of the price rise were fossil fuel–fired (natural gas) power producers—whose emissions accelerate climate warming (Hiar, 2021).

Prospects of change. The hydropower projects that have received most attention recently are the small ones. In 2018, for example, the Alaska Native village of Igiugig applied to the Federal Energy Regulatory Commission (FERC) for a license to begin a pilot project using river power technology that would lower diesel use in the community and reduce electricity costs. The "RivGen power system" technology would place two turbines in the current supported by pontoons. No dams or barges would be used, so there would be no barrier to obstruct fishing or river traffic (Hobson, 2018). The second project has been in the same family for nearly 100 years and operates three small hydro plants north of New York City (on the Hudson River) (*E&E News*, 2019). These small projects lack the economy in scale of large hydro or gas turbines, but have offsetting advantages—fewer adverse environmental effects and more engagement of small community values.

The Biden administration's focus on renewables improves prospects for hydropower in the United States. For example, researchers at the DOE's National Renewable Energy Laboratory believe that linking floating solar panels with hydropower might produce up to 40 percent of the world's energy in the future. The researchers argued that constructing solar panels on the surface of hydro reservoirs and feeding the power they produced into the same substation would reduce costs of both energy resources and increase their reliability (by capitalizing on synergies of solar and hydro operations; see Iaconangelo, 2020).

Democratic senator Maria Cantwell of Washington and Republican senator Lisa Murkowski of Alaska have taken advantage of Biden's focus on clean energy to address long-standing problems with hydropower infrastructure. They introduced in Congress the Maintaining and Enhancing Hydroelectric and River Restoration Act of 2021, to create a 30 percent federal tax incentive to spur investments in dam safety, grid flexibility, and environmental improvements. Murkowski, representing a state in which hydropower provides 25 percent of the electricity, noted that hydro also could provide "black start" capabilities (the ability to bring electric power stations and other grid components back online without having to use an external electric power transmission network) to assist recovery from blackouts (*Hydro Review*, 2021). Finally, pumped storage comprises a growing share of hydropower projects under development or proposed, and this would likely promote renewables (Willson, 2022).

Still, the message from both government and industry is that hydropower is valued in the current energy mix, but it must compete with both the old (fossil fuel) sources of energy and the newly emphasized renewables.

FURTHER READING

Bracmort, Kelsi, Adam Vann, and Charles Stern. *Hydropower: Federal and Nonfederal Investment*. Congressional Research Service (CRS), July 7, 2015.

E&E News. "N.Y. communities try tiny hydro plants," July 9, 2019.

Hiar, Corbin. "Hydropower withers in drought, boosting fossil generators," *E&E News*, August 19, 2021.

Hobson, Margaret. "Diesel-dependent Alaska town files with FERC for hydro project," *E&E News*, December 5, 2018.

Hydro Review. "U.S. Senators introduce bill to incentivize hydropower," June 29, 2021. https://www.hydroreview.com/regulation-and-policy

Iaconangelo, David. "DOE study: Solar-hydro projects could power 40% of world," *E&E News*, October 2, 2020.

Jacobs, Jeremy. "Bonneville power dynasty: No longer a 'no-brainer,'" *E&E News*, November 27, 2019a.

Jacobs, Jeremy. "Near disaster revealed nationwide dam problems," *E&E News*, December 30, 2019b.

Reisser, Wesley and Colin Reisser. *Energy Resources: From Science to Society*. New York: Oxford University Press, 2019.

Reuters. "Droughts shrink hydropower, pose risk to clean energy," *New York Post*, August 13, 2021.

U.S. Bureau of Reclamation (USBR). "Hydropower program," updated March 2, 2016. https://www.usbr.gov/power/edu/hydrole.html

U.S. Department of Energy (DOE), Energy Efficiency & Renewable Energy (EERE) Office. *2014 Hydropower Market Report Highlights*, April 2015. https://energy.gov/eere/water/downloads/2014-hydropower-market-report

U.S. Department of Energy (DOE), Energy Efficiency & Renewable Energy (EERE) Office. *U.S. Hydropower Market Report*, January 2021. https://www.energy.gov/eere/water/downloads/us-hydropower-market-report

U.S. Department of Energy (DOE), Energy Efficiency & Renewable Energy (EERE) Office (WPTO). *Hydropower Vision: A New Chapter for America's 1st Renewable Electricity Source*, July 26, 2016. https://www.energy.gov/eere/water/articles/hydropower-vision-new-chapter-america-s-1st-renewable-electricity-source

Willson, Miranda. "Public outcry drains pumped hydro's 'watershed moment,'" *E&E News*, May 7, 2021.

Willson, Miranda. "Pumped storage is having a moment. Will it shift renewables?" *Energywire*, April 15, 2022.

Q11. IS IT FEASIBLE TO DOUBLE THE AMOUNT OF NUCLEAR POWER IN THE U.S. ENERGY MIX IN THE MID- TO LONG TERM (25–50 YEARS)?

Answer: Yes, but substantial government and private-sector investment would be needed.

The Facts: Nuclear fission is a process in which the nucleus of an atom is split into two smaller (and lighter) nuclei. The process releases a great amount of energy (splitting apart atoms to form smaller ones), producing heat and steam that turn turbines to generate electricity in nuclear power plants. Nuclear fission is a very fast and efficient way to create energy, and it does so without producing any carbon (no greenhouse gases).

Status of nuclear power in 2022. The United States generates more nuclear power than any other nation; in 2020, it comprised 20 percent of America's electricity (exceeding coal's share of energy production for the first time in history). On December 31, 2021, a total of 93 nuclear reactors operated in 55 nuclear power plants, distributed in 28 states with fewer plants in the West than in other regions (EIA, 2022). Two new nuclear reactors are scheduled to be completed within a few years; the last reactor to close was New York's Indian River plant in April 2021. (Exelon Corp. had threatened to close two plants in Illinois because their operations were unprofitable, but the state agreed to a $700 million subsidy [Gardner, 2021].)

Nuclear power plants are owned by private corporations but extensively regulated by the government, including onsite inspection by the Nuclear Regulatory Commission (NRC, 2020). The Department of Defense (DOD) manages nuclear stockpiles for military purposes; the Department of Energy (DOE) operates 21 national laboratories (usually managed by private-sector firms), several of which engage in nuclear energy research. The degree of private ownership of civilian nuclear facilities is unusual globally, as is the high degree of private–public-sector cooperation and coordination.

Aging power plant infrastructure. Given the carbon-free qualities of nuclear energy, the technology seems an excellent candidate for expansion in some respects. However, a number of obstacles stand in the way of a reinvigoration of the industry, starting with its aging infrastructure. The average age of nuclear plants is 39 years, but most nuclear plants use steam generators that are required to be replaced after 20–30 years of service. The replacement costs owners as much as several hundred million dollars per reactor, so the expense looms as a significant obstacle to modernization (CRS, 2016).

Nuclear waste disposal. There are several kinds of nuclear waste, from highly radioactive material to low-level sludge (the bulk of all waste, which has been exposed to little more than 1 percent of radiation). The highly radioactive waste (called *radionuclides*) have excess nuclear energy, making them unstable; they have half-lives (the time taken for half of their atoms to decay) extending billions of years. Most of the highly radioactive kind of nuclear waste is placed deep underground, in geologically inactive zones that are not at risk of earthquake or other destabilizing natural forces. Nuclear waste is found at 131 sites in 39 states, waiting for final disposal. Congress approved the Yucca Mountain facility in Nevada as the central repository for civilian plant waste. However, criticism of the selection of Yucca Mountain, an area just 80 miles from Las Vegas, was strong from many Nevadans, as well as antinuclear groups nationwide. In 2009, the Obama administration suspended the transfer (Cassedy & Grossman, 2017). Initially, President Trump reinstated the Yucca Mountain site as a repository, but then reversed course after the announcement sparked a political backlash in Nevada.

Division in public perceptions. Public opinion on nuclear power generally has been positive. In the early 2000s, polls showed three times as many people strongly supporting nuclear energy as opposed to it; support was firm also for recycling of nuclear fuel. When asked whether respondents approved a diversified electricity mix, a large majority agreed too. However, nuclear power's reputation suffered in the aftermath of the second-worst nuclear accident in history in March 2011. The incident occurred in Fukushima, Japan, after a severe earthquake (nearly 9 on the Richter scale) struck off Japan's eastern coast, causing a powerful 49-foot tsunami.

Although reactors at the Fukushima facility were shut down instantly, all three cores melted, and radiation leaked out of the plant into the immediate environment. After this incident, only 40 percent of a sample of American adults believed that benefits of nuclear power outweighed its risks (CRS, 2016). A 2019 Gallup poll found the public divided. While 49 percent of Americans surveyed favored nuclear power, the rest did not. Moreover, attitudes divided along party lines: 65 percent of Republicans favored nuclear power use; only 42 percent of Democrats did (Brugger, 2019). This division makes increasing public investment in nuclear expansion challenging.

Investment in large-scale nuclear power plants. Although some owners closed large-scale plants in order to avoid paying high costs to replace generators, most of the retirements appear to be for other reasons. The 2016 study by the Congressional Research Service (CRS) found that most nuclear power plants considered vulnerable to shutdown before the expiration of their operating licenses were "merchant plants," operating in a competitive market. The minority of plants operated in an environment where the cost of electricity was based on regulator-approved charges, operating expenses, and a reasonable investment return. For most plants, CRS analysts believed that government needed to act to provide tax incentives, or establish a carbon price (making nuclear-based electricity less expensive than that of fossil fuels), or set up power purchase agreements with federal agencies over multiple years. Such capital investments have not yet been made.

There are economies of scale associated with very large nuclear plants. They are big enough to afford specialized personnel to prepare detailed consolidated permit requests to government agencies. Their size gives them clout in negotiating with third-party vendors and with labor unions as well. However, these advantages did little to facilitate construction of Plant Vogtle, the nuclear power plant designed by Westinghouse before it went bankrupt in 2017. It is still under construction in Georgia but is now owned by Southern Co. Five years behind schedule in 2022, it is likely to cost more than twice its original $14 billion estimated budget. Plant Vogtle is the first set of reactors to be built in the United States in 30 years, but the soonest it is likely to open is the first quarter of 2023 (Williams, 2022). Once completed, forecasters believe, Plant Vogtle's four units will provide electricity for 1 million Georgia homes and businesses (as well as train thousands of workers in nuclear plant construction).

Investment in small-scale plants. Prospects look better for investment in moderate to small modular nuclear reactors (SMRs). These reactors are still at the design and experimentation stage, but they appeal to both regulators and business firms for several reasons. Specifications limit the size of a group of reactors to 2,400 MW or 800 MW for each individual reactor, meaning they are not likely to ever experience a blowout. Their smaller size also means they can be built in a factory setting, thus avoiding giant cost overruns afflicting large-scale plants. SMR designs occupy less land and are purported to have broader safety benefits, which advocates claim will increase local-level support for these facilities. Minimizing the emergency evacuation zone for residents and workers around nuclear plants (the so-called "plume zone") would also decrease operator costs, making the nuclear fuel better able to compete with low-cost electricity from producers of natural gas and renewables (Anchondo, 2019).

Researchers at the Pacific Northwest National Lab and MIT compared how the leading SMR (being developed by NuScale Power) would compete with three competitors in a world with nearly no carbon: natural gas, wind plus a battery system, and electricity from natural geothermal sources. They found that the geothermal model came in first, followed by the NuScale SMR, with the wind-battery combination coming in last (Behr, 2021a). However, like large-scale nuclear plants, next-generation reactors await technological advances; they require stable partnerships with industry and federal government assistance (Swartz, 2021a).

Trump administration nuclear ventures. In 2017, President Trump told industry executives that he wanted to "revive and expand our nuclear energy sector . . . which produces clean, renewable and emissions-free energy." Initially, members of the Congress were not involved in policy discussions, but Congress did enact legislation in 2018 requiring the NRC to develop a regulatory framework, and to revise its budget process and fee program—all of which were done in 2019 (Northey & Dillon, 2019).

Trump took more actions to benefit the nuclear power sector than previous Republican presidents. The first Trump energy secretary, Rick Perry, was a tireless advocate of nuclear power. During his first year in Trump's cabinet, Perry proposed to bolster faltering coal and nuclear plants (which had fuel available on-site) by emphasizing that they had met the definition of reliability and grid resilience—the ability of the electricity system to recover quickly from disruption. This approach was roundly denounced by members and advocates of the renewable energy industry. The FERC under Neil Chatterjee's leadership rejected Perry's proposal and sided with the renewables against nuclear (Dillon, 2021).

A reinvigoration of domestic uranium production was another area of emphasis in the Trump administration, which satisfied both advocates of the U.S. mining industry and critics of state-owned enterprises (mostly of China), whose "coercion" and "predatory economics" threatened U.S. national security (Marshall & Dillon, 2020). (Only 13 percent of U.S. uranium imports come from Russia and China; most originate in Canada, Kazakhstan, and Australia.) One product of the Nuclear Fuel Working Group, formed by President Trump, was a budget proposal to build a $1.5 billion uranium reserve over 10 years, beginning with a purchase of uranium from U.S. mines, using domestic conversion and enrichment services. The plan would expand access of mining companies to high-quality uranium deposits on federal lands in the West. The omnibus spending bill passed by Congress in December 2020 included $7.5 million of initial funding for this research.

Biden administration changes. Joe Biden was different from the other 2020 presidential election campaigners because he stressed the need to develop strategies of nuclear power emphasizing innovation. Like most of the other Democratic candidates, he mentioned the need to develop renewables as a primary means to address climate warming; unlike most, his clean energy platform had room for nuclear, and he was not averse to calling for a new generation of reactors more efficient than those already in place (Klump & Swartz, 2021).

Additional differences between the Biden and Trump administrations concerned the undergirding of nuclear power development. First, from the outset, nuclear power has been advertised as central to a national Clean Energy Standard (CES). As Biden administration senior climate advisor Gina McCarthy explained, such a CES was critical for the United States to reach zero-carbon emissions (Clark, 2021); however, it was cut from the budget proposal to reduce opposition from Republicans and industry allies (see Q9). Second, the Biden administration's response to challenges confronting the nuclear industry is to refocus on investment in nuclear as part of the national $2 trillion infrastructural development proposal.

Supporting this approach is PJM Interconnection, which oversees the power grid in 13 Midwestern and mid-Atlantic states and has been troubled by the financial straits of 16 nuclear plants in the region. None of the plants is near retirement, but all suffer from underinvestment as compared to coal and natural gas. The Nuclear Energy Institute (NEI), the trade group for the nuclear power industry, argues that costs for nuclear plants are high because they reflect commitments to operate three years into the future, and unlike fossil fuels cannot take advantage of quick market fluctuations in demand for their product (Tomich, 2021).

A third difference has been the willingness of the Biden administration to seek approval from nuclear oversight agencies (the NRC, primarily) for extension of permits for operation of nuclear reactors to 80 years. After energy giant Dominion received approval from the NRC for a permit extension for a nuclear facility in Virginia, the state's governor praised the decision: "Carbon free, around the clock nuclear power and the well-paying clean energy jobs it creates is a vital part of achieving that goal (of a carbon free electricity grid)" (Swartz, 2021b). However, after a legal challenge by environmental and consumer groups, the NRC instructed companies applying for 20-year extensions that they must undergo NEPA reviews of their aging equipment (Swartz & Dillon, 2022).

Overall, the Biden approach demonstrates willingness to give budget priority to nuclear renewal and advanced reactor research. The Nuclear Energy Institute (NEI), the industry's major trade association, included among its proposals for the Biden infrastructure plan a credit program to bolster at-risk nuclear reactors from closing prematurely. The program includes $6 billion to help struggling nuclear plants remain competitive with less expensive natural gas (Cahlink & Dillon, 2021). The NEI also recommended that the Biden administration assist the development of advance reactors, the latest stage of SMR evolution that represents a pragmatic approach to the "safe enough" nuclear reactor debate. (Proponents of the safe enough perspective hold that making reactors perfectly secure against all conceivable threats or scenarios would make them unaffordable; an achievable goal is to build SMRs that protect the public and environment against plausible emergencies and accidents.) Energy Secretary Granholm pledged $700 million in support of advanced reactors as evidence of the Biden administration's commitment to clean energy (Behr, 2021b).

In 2021, the U.S. Geological Survey (USGS) proposed removing uranium from the national list of critical minerals, arguing that the Energy Act of 2020 explicitly excluded minerals related to fuel from being defined as critical minerals (Holzman, 2021). However, because nuclear provides carbon-free electricity essential to an effective climate warming strategy, it remains on the agenda of the Biden administration. Its Energy Department prepared its own list of critical elements, calling uranium a "material of concern," leaving the door open for its inclusion in the future (Holzman, 2022). The Russian invasion of the Ukraine in February 2022 introduced a new area of complexity—its control of Europe's largest nuclear plant (Zaporizhzhia, with six reactors, in the Ukraine) and wartime conditions' effects on cost and flow of fuel [low-enriched uranium] to existing advanced U.S. reactors (Northey, 2022).

The most recent report of the International Energy Administration (IEA) puts nuclear (and hydropower) in perspective. They "provide an essential foundation for transitions" to a decarbonized grid, and would remain essential by 2050 (Waldman, 2021). Yet persistent issues related to cost of development, limited political support, and environmental safety dim prospects for nuclear's ability to compete with renewables, particularly wind and solar, in the years ahead.

FURTHER READING

Anchondo, Carlos. "Nation's first small reactor project moves forward," *E&E News*, July 22, 2019.

Behr, Peter. "DOE lab: Small reactors beat wind, batteries in future grid," *E&E News*, May 28, 2021a.

Behr, Peter. "NRC historian details 'safe enough' reactor debate," *E&E News*, June 21, 2021b.

Brugger, Kelsey. "Americans are split on nuclear power. Why?" *E&E News*, March 28, 2019.

Cahlink, George and Jeremy Dillon. "Energy winners and losers in the bipartisan infrastructure package," *E&E News*, August 10, 2021.

Cassedy, Edward and Peter Grossman. *Introduction to Energy: Resources, Technology, and Society*, 3rd ed. New York: Cambridge University Press, 2017.

Clark, Lesley. "Gina McCarthy: Clean energy standard to include nuclear, CCS," *E&E News*, April 2, 2021.

Congressional Research Service (CRS). *Financial Challenges of Operating Nuclear Power Plants in the United States*, updated December 14, 2016. https://www.everycrsreport.com/reports/R44715.html

Dillon, Jeremy. "Chatterjee on coal, nuclear push: 'I didn't handle it well,'" *E&E News*, July 28, 2021.

Gardner, Timothy. "Illinois approves $700 million in subsidies to Exelon, prevents nuclear plant closures," *Reuters*, September 13, 2021.

Holzman, Jael. "USGS proposal yanks uranium from critical minerals list," *E&E News*, November 9, 2021.

Holzman, Jael. "Uranium may regain 'critical' status despite USGS move," *Greenwire*, March 1, 2022.

Klump, Edward and Kristi Swartz. "Miss. plant raises concerns about nuclear power," *E&E News*, January 6, 2021.

Marshall, James and Jeremy Dillon. "Report lays out plans to save sagging U.S. nuclear industry," *E&E News*, April 23, 2020.

Northey, Hannah. "How Russia's invasion is affecting U.S. nuclear," *Energywire*, March 14, 2002.

Northey, Hannah and Jeremy Dillon. "Trump's nuclear revival? It's a 'black box,'" *E&E News*, January 15, 2019.

Swartz, Kristi. "'Something's wrong.' Vogtle inspection raises concerns," *E&E News*, August 25, 2021a.

Swartz, Kristi. "Dominion wins approval for 20-year reactor extension," *E&E News*, May 5, 2021b.

Swartz, Kristi and Jeremy Dillon. "Feds walk back plans for nuclear reactors to run 80 years," *Energywire*, February 25, 2022.

Tomich, Jeffrey. "Report shows financial troubles of 16 U.S. nuclear plants," *E&E News*, March 12, 2021.

U.S. Department of Energy, Energy Information Administration (EIA). "FAQs: How many nuclear power plants are in the United States, and where are they located?" March 7, 2022. https://www.eia.gov/tools/faqs/faq.php?id=207&t=21

U.S. Nuclear Regulatory Commission (NRC). "Backgrounder on oversight of nuclear power plants," January 2020. https://www.nrc.gov/reading-rm/doc-collections/fact-sheets/oversight.html

Waldman, Scott. "Gina McCarthy: Nuclear energy isn't going anywhere," *E&E News*, May 20, 2021.

Williams, Dave. "Plant Vogtle nuclear projects hits another delay with 'construction quality' issues," *The Augusta Chronicle*, December 3, 2021.

Q12. IS DEVELOPMENT OF WIND AND SOLAR POWER LAGGING BECAUSE OF AN INSUFFICIENT PRODUCTION OF METALS (E.G., COBALT, CADMIUM, LITHIUM, AND RARE EARTH ELEMENTS) USED IN THOSE TECHNOLOGIES?

Answer: No, but other factors have slowed their development.

The Facts: The amount of solar energy absorbed by the earth every year is enormous, and energy from the sun has been used from time immemorial. Recent use of solar heating and energy is, however, a significant enhancement of such efforts in the past. Meanwhile, wind power has propelled vessels for millennia, and windmills have pumped water for centuries; only since the late nineteenth century, though, has wind force been refined sufficiently to generate electricity.

Current status of wind/solar production. In April 2019, daily electricity generation from wind turbines reached a high of 1.42 million megawatt hours (MWh), a record not surpassed until December 2020. The United States is the second-largest wind power market in the world, and on average wind accounted for 9 percent of U.S. electricity generation in 2020 (EIA, 2022b). It delivers more than a fifth of the electricity produced in the six states of Kansas, Iowa, Oklahoma, North Dakota, South Dakota, and Maine (AWEA, 2019). Solar energy enjoyed a boom in 2020, capturing a large share of new electric-generating capacity (Fialka, 2020). From only .34 gigawatts (GW) in 2008, it grew to 97.2 GW in 2021. Today, more than 3 percent of U.S. electricity comes from the sun (EERE, 2021), through two different means. The photovoltaic method directly converts sunlight, through silicon panels, into electricity, whereas the concentrating solar power (CSP) method uses steam turbines to generate electricity. Unlike wind, which has experienced a steady increase of consumption, the solar power industry has experienced rapid spurts of growth, as well as dramatic reversals (attributable to policy reforms in nonresidential markets of a few states). The future status of solar investment tax credits is another factor (SEIA, 2019).

The critical (and scarce) metals issue. The factors needed to generate renewable energy include photovoltaics, fuel cells, light turbines (used for wind power generation of electricity), and storage batteries with greatly increased capacity compared to those in use. The raw materials of the industry include common minerals such as copper and zinc. Newer processes used in photovoltaics, in cell phones, and some storage batteries employ other minerals found infrequently in nature, called "rare earth elements" (REEs). They include 17 elements, 10 of which are used more than others (see Alonso et al., 2012).

For several reasons, the growth in demand for renewables seems likely to outstrip available supply of raw materials and equipment. The first concern is uncertainty of production increases and unsustainability of mining practices to produce the minerals. A 2021 report of the International Energy Administration (IEA) found that continued shortages of lithium loomed as a potentially significant hurdle in creating sufficient supply of batteries used to power electric vehicles (EVs; see Q7). In the absence of new lithium mining projects, lithium prices are projected to triple and EV battery prices to increase by 3–4 percent (Willson, 2021b).

The 2021 annual renewables report stated that continued high prices of key raw materials were changing the economics of clean energy technologies (Iaconangelo, 2021b). At the onset of the Ukraine war in February 2022, observers feared that rising costs would stymie further expansion (Iaconangelo, 2022).

The U.S. mining industry contends that extraction activities for rare earth elements are technologically challenging but do not cause extensive environmental damage (Ferris, 2021). However, environmental organizations strongly oppose continued exploration/development of REEs and other renewables' mining projects. For example, Basin and Range Watch and other green groups sued BLM in U.S. District Court of Nevada to stop Lithium Americas Corp.'s proposed Thacker Pass project. The objection was based on the assertion that an existing mine in the same area—the Silver Peak mine—already threatened the habitat of the endangered greater sage grouse (Marshall, 2021b).

A second concern is that countries with near monopolistic control of raw materials used in renewables production may ban sales to the United States. In 2021, about 70 percent of cobalt was produced by the Congo, about 60 percent of REEs' production was controlled by China, and Australia produced more than half of the global supply (Marshall, 2021). Under the America First agenda of the Trump administration, U.S.-China trade relations were particularly tense, as the president complained that China was largely responsible for the massive U.S. trade deficit. In 2018, the president raised U.S. tariffs on foreign solar panels and cells by 30 percent and then a few months later assessed another 25 percent tariff on solar equipment from China (Iaconangelo, 2020b).

The Chinese retaliated by raising tariffs on U.S. agricultural imports. The stage 1 trade agreement of the United States and China did not allay these concerns. In June 2021, the Biden administration banned some solar imports from China because of alleged human rights abuses in Xinjiang province, in which 40 to 50 percent of global polysilicon, used in most solar cells, is produced (Ferris & Iaconangelo, 2021). The Russian invasion of Ukraine presented a new crimp in supply of metals essential to clean technology, which slowed the energy transition (Holzman & Storrow, 2022).

Democratic and Republican leaders of the Senate Energy & Natural Resources Committee developed an American Mineral Security Act to reduce U.S. dependence on foreign suppliers (Beitsch & Frazin, 2020), but Senate and House efforts languished in Congress. The Biden administration has taken steps, however, to support REE procurement. For example, it announced $18 million in grants to improve the domestic mineral supply, such as extracting REEs from coal and coal waste (Marshall, 2021a).

A third factor making future prospects of critical minerals uncertain is the unpredictable nature of technological innovation. Recycling of batteries has been slow to develop, even as the sheer volume of old batteries has increased. An Earthworks report of April 2021 predicts that "effectively recycling end-of-life batteries could reduce global EV mineral demand 55% for newly mined copper, 25% for lithium and 35% for cobalt and nickel by 2040" (Earthworks, 2021). Commercial mining of sea bed nodules of critical and scarce minerals is several years away from being feasible without serious environmental degradation, however.

Finally, even were the technological and other problems overcome and large amounts of critical minerals excavated, "the majority would have to be shipped abroad for processing," according to Abigail Wulf, director of the Center for Critical Minerals Strategy for SAFE-Securing America's Energy Future, a nonprofit and nonpartisan advocacy group seeking to reduce America's dependence on fossil fuels (Marshall, 2021d). Mining operations in the emerging economies are less constrained by environmental regulations than are those in the United States. Of course, the same applies to refining of materials such as lithium and nickel, for people in the neighborhood of refineries and smelters will protest and often prevail against the operations. Thus, what is called a "race to the bottom" occurs, in which U.S. environmental pollution is exported to (or condoned in) emerging economies overseas, such as those in China and the Congo. Still, other factors inhibit wind/solar power development to a greater degree than insufficiency of critical minerals.

Infrastructure needs. Because wind and solar power are relatively recent additions to the energy mix of the United States, essential infrastructure to bring the power they produce to market is needed. Buildings and other needs specific to locations are greater for wind than solar. For solar, the primary infrastructure needs are places to dispose of materials with toxic chemicals, such as solar panels and batteries. Installing wind farms requires construction of turbines and their repair and resupply. The infrastructure needs are those accommodating vehicular traffic, such as roads for transport of heavy components such as rotors, blades, and towers (Iaconangelo, 2019b).

An additional challenge is presented by wind farms that are offshore. More than a dozen such wind farms have been proposed off the Northeast coast, and the Bureau of Ocean Energy Management (BOEM) estimates that 2,000 turbines will eventually be erected, likely from large floating platforms 20 to 40 miles off the coast (Richards, 2020c). Many of the facilities such as warehouses and storage yards that are needed to supply offshore wind to mainland customers will be constructed onshore, which in the next decade will create new opportunities for oil field service firms. In addition to establishing warehouses, their ships will be needed to install turbines, and their fleets of barges and tugs will move equipment and employees from shore to offshore platforms (Lee, 2021).

Once wind or solar equipment is installed, transmission lines need to be set up. These are called transmission "backbones" (see chapter 4, too), meaning they are high-speed transmission lines connecting the wind or solar source to the market some distance away. Wind operators such as Equinor ASA and grid operators like PJM Interconnection comment that extensive collaboration between states that regulate utilities and planning authorities (the major regional transmission organizations or RTOs) will be needed to make such investments possible. Grid connection problems are an overlooked issue in clean energy development (Willson, 2022).

Moreover, many energy development rules currently in force were originally designed for coal and nuclear plants. As legacy plants are taken offline, the places they vacate are sought by wind and solar generators, and interconnection to the grid may be worth hundreds of millions of dollars (Tomich, 2021). Yet rules have yet to be developed to allow them to compete fairly in the utilities market.

Private investment and public (tax) incentives. The November 2020 election of Joe Biden and his vigorous climate change agenda changed the dynamics of corporate investment in renewables. Even the fossil fuel industry has taken note. Chevron has supported commercialization of floating offshore wind turbine technology (Anchondo, 2021a), and BP's CEO Bernard Looney announced investments in hydrogen-based green technology patterned after previous investments in offshore wind (Anchondo, 2021b). Finally, the activist investment firm Engine No. 1 (with a clean energy agenda) succeeded in placing two of its four candidates on ExxonMobil's corporate board (Deveau et al., 2021).

Renewables, and particularly wind and solar, have greatly benefitted from loan guarantees, tax breaks (for producers and consumers), and special grants. One incentive, the investment tax credit (ITC), can be claimed by solar companies indefinitely for investment in utility-oriented operation (the tax credit for residential systems is being phased out). It also has been available for offshore wind, to assist in reducing upfront costs (Iaconangelo, 2019c). Production tax credits also have been available for wind and solar with varying terms (EIA, 2022a). Congressional support of tax incentives for renewables has been bipartisan, reflecting the fact that the industry is growing in so-called blue (Democrat-leaning) and red (Republican-leaning) states alike.

A prominent think tank called the Information Technology & Innovation Foundation (ITIF) asserted in a 2019 report that federal tax credits for solar and wind power should be reformed. The report's authors remarked that existing credit extension policies favor mature technologies over younger, currently more expensive alternatives that could over time prove to be superior. The ITIF believes that both wind and solar industries will continue to grow, and it strongly supports providing tax credits to early adopters of new technologies and innovations (ITIF, 2019).

Social and environmental issues. Wind power receives more criticism on social and environmental grounds than solar. A few scientific studies have suggested that heavy reliance on wind power could increase local surface temperatures, because turbines mix the air near the surface, redistributing heat. Critics of the studies respond that their finding does not apply to warming of the seas (a large concern of climate change research), and that the overall adverse warming effects will be "extremely minor" (Marshall, 2018). A related question is whether large amounts of solar power on the grid cause an increase in air pollution and greenhouse gas emissions. A study by Carnegie Mellon researchers found the increase of emissions mainly was of nitrous oxides (NOx), and it did not counteract the overall positive effect of using combustion turbine power plants in conjunction with solar. Both wind and solar on the grid were found to have negligible effects on carbon emissions (Swartz, 2019).

Birds and fish are vulnerable to installations of wind turbines, and the subject of greater attention. Regulatory agencies such as the Army Corps of Engineers and the Energy Department have worked with the Sierra Club and Environmental Defense Fund to ensure that wind farm developments on Lake Erie meet NEPA guidelines (which explicitly state that the project should "have no significant impact to birds or on the environment"), but this did not deter two other environmental organizations from suing the agencies in federal district court. They alleged that one potentially vulnerable bird species, the Kirtland warbler, ran the risk of being "diced in a wind turbine" (Farah, 2019).

Two scholars collected perceptions of commercial and recreational fishers on potential ecological impacts of the Block Island Wind Farm (BIWF), the first such farm built in the United States. In general, the respondents appreciated the fact that the wind turbines created a new structure for fish habitat, an "artificial reef." This changed the trophic structure (spreading fish along the tower) and increased the abundance of fish (Ten Brink & Dalton, 2018). Fishermen's groups have been concerned about navigating among wind turbines, though. One alliance of groups objected to the proposed one nautical mile distance requirement between turbines in the New England Outer Continental Shelf (OCS), recommending instead that the turbines be spaced at least four nautical miles apart to reduce the risk of accidents at sea, whether between vessels or between vessels and turbine facilities (Richards, 2020a).

The authorizing and regulatory federal agency for offshore energy facilities, the Bureau of Ocean Energy Management (BOEM), completed a cumulative analysis of impacts of offshore wind power development. The agency report concluded that offshore wind farms might have a major "adverse" impact on some commercial fisheries, marine traffic, and the Defense Department. Responding to the report, the Responsible Offshore Development Alliance (RODA) called for a five-year moratorium on the offshore wind business, a decision supported by the Department of the Interior's (DOI) solicitor in late 2020 (Richards & Doyle, 2020). Janet Mills, the Democratic governor of Maine, had signed legislation permanently banning offshore wind from state waters, primarily to protect lobster fishing. As a proponent of offshore wind, she received praise for her careful planning to find a place in the Gulf of Maine where an array of wind turbines would have a minimal impact on fishing (Richards, 2021d).

Wind power projects appear likely to draw more concerted and widespread opposition when they are located relatively close to shore. Conversely, wind power projects farther out at sea have typically drawn fewer challenges. One federal study of East Coast tourists found that most would not be opposed to utility-scale wind turbines if they were 20 miles from shore; respondents saw fewer negative impacts than at 2.5 miles from shore (Iaconangelo, 2019a). The chief author of the study, Jeremy Firestone, noted that when given a choice, poll respondents were more likely to prefer living close to wind turbines as compared to power plants using fossil fuels (Iaconangelo, 2019d). Observers wonder, however, whether future OCS wind farms will face as much controversy and resistance as several pilot projects off the Massachusetts coast have experienced. In these cases, complaints have been leveled for allegedly detracting from the aesthetic beauty of the Atlantic coastline.

Biden administration changes. Three large changes distinguished the Biden approach from the Trump administration's skepticism of solar and wind. First, climate change was a driver of change in a "whole of government" strategy to lowering carbon dioxide emissions. Greg Wetstone, CEO of the American Council on Renewable Energy (ACORE), said: "We've been working for years to prepare to support exactly this kind of agenda" (*E&E News*, 2021a). This enthusiasm for wind and solar was reflected in President

Biden's first budget request to the Congress. The proposal called for investing millions of dollars in support of wind power. It also called for increased funding for renewables in the budgets of important energy-related agencies in the federal government, including the departments of Interior and Energy and the Environmental Protection Agency (*E&E News*, 2021b).

Second, to assert U.S. leadership on clean energy technology, Biden signed an executive order to assess supply chain vulnerabilities in electric vehicle batteries and critical minerals including REEs. The order also included semiconductors because of auto industry complaints about shortages; pharmaceutics was included as well because of shortages of personal protective equipment across much of the United States when the COVID pandemic emerged in March 2020 (Marshall, 2021c).

Third, the new administration made significant advances with several proposed offshore wind farms, in pursuit of its climate objectives. Mentioned above was the Block Island Wind Farm, actually a pilot project developed off the Rhode Island coast in 2018. The environmental impact statement (EIS) was completed for the first full-scale wind farm (called the Vineyard Project as it has been developed off the coast of Martha's Vineyard, Massachusetts) (Richards, 2021b). When completed in 2023, its 62 turbines are expected to power 400,000 homes and generate 3,600 jobs. BOEM in 2021 reviewed smaller project for 15 turbines called "South Fork," off the Rhode Island coast. Another proposal called the "Revolution Wind Farm" project off the coasts of Rhode Island and Connecticut restarted the EIS process in 2021 after stalling out during the Trump administration (Richards, 2021c). Later in 2021, BOEM began EIS review of the South Wind proposal off the coast of New Jersey, a development that proponents say will power 500,000 homes. All of these efforts are taking place as the Biden administration also seeks to reduce time spent in the permitting process, expand additional projects to a total of 16 by 2025, reach accommodations with commercial fishing interests, and offset 78 million metric tons of carbon dioxide (Richards, 2021a). Before the Trump-era moratorium on wind leasing in the south Atlantic took effect, the Biden administration scheduled a major lease sale for May 2022, but it was uncertain whether Duke Energy Corp.—the largest North Carolina utility—and utility regulators were committed to offshore wind (Richards & Swartz, 2022).

Future prospects. The near- and midterm prospects for both wind and solar seem good, especially because their costs are expected to decline vis-à-vis other energy sources. NextEra Energy, the global leader in renewable energy development, forecasts unprecedented growth under President Biden's climate agenda, which seeks to accelerate the nation's shift to wind, solar, and other renewables (Swartz, 2021).

Technological breakthroughs such as increased lithium battery storage have sparked private-sector interest, as noted in the new wind projects. Perhaps the brightest spot for renewables was a scientific breakthrough for solar. Engineers from three countries met performance requirements for an advanced solar cell. Unlike silicon photovoltaics, which are mounted in an array on rooftops or in fields, the new solar cell (called a *perovskite*) is pliable and can be ingrained into windows or sprayed on walls (Iaconangelo, 2020a). If perfected, this technology could bring about a significant increase in low-cost, clean energy (Behr, 2021; Klump, 2020).

States play a large role in the expansion of renewables, as noted in the efforts of wind power firms to meet state utility commission regulations. California has the nation's most ambitious renewables mandate and must meet carbon neutrality by 2045. It is in the process of speeding up its permitting process to help companies engaged in solar, wind, battery storage, and other green technologies. A report of the California Public Utilities Commission, Air Resources Board and Energy Commission stated, "California will need to sustain its expansion of clean electricity generation capacity at a record-breaking rate for the next 25 years." California is not alone in shifting state policies to support renewables; 30 states have set mandates for different amounts of low-carbon power (Mulkern, 2021).

Comparing solar and wind expense for consumers. Several studies have suggested that rooftop solar appears to benefit the wealthy over the poor (who are less able to afford installation costs). They point out that the system measuring usage of electricity in California, called "net metering," is effectively a subsidy to solar by other ratepayers. The Solar Energy Industries Association (SEIA) proposes using a "net billing" system instead, which would better align the benefits of rooftop solar to grid needs (Willson, 2021a).

Net billing means billing customers for retail electricity purchased at retail rates, while crediting customers' bills for any customer-generated electricity (e.g., rooftop solar) sold to the utility. Net metering is a billing system that credits solar energy system owners for the electricity they add to the grid. The argument is that solar system owners gain credits, while non-solar customers (ratepayers) subsidize those getting the credits.

A related dimension to the future of solar in the United States is the changing dynamics of trade between the United States and China. In late August 2021, a new coalition of solar companies cautioned that "exploitative" Chinese trade practices endangered U.S. clean energy goals. Calling themselves the American Solar Manufacturers Against Chinese Circumvention (A-SMACC), they requested that the Biden administration impose new tariffs on solar imports from China that were being routed through Malaysia, Vietnam, and Thailand (to avoid paying U.S. tariffs). They filed petitions with the Commerce Department, publicly backed by at least three companies, to which SEIA objected because the trade association feared the impact of the petitions would disrupt the domestic solar market (Iaconangelo, 2021a). The changing state of U.S.-China relations goes beyond the scope of this volume, but is mentioned here to emphasize how trade tensions and policies could impact availability of essential solar supplies for U.S. manufacturing as well as the evolution of trade associations as renewables gain importance in U.S. energy policy.

These uncertainties about the cost of solar are also an issue with wind costs. The point concerned the reliability of estimates in the falling price of wind as compared to coal in the last decade or two. A team of researchers led by scientists at the Lawrence Berkeley National Laboratory conducted detailed surveys with 140 wind experts, who tended to underestimate the technology's advances on coal. The report recommended incorporating uncertainties about cost into planning models, and "avoid putting all your eggs in the same basket." Senior report author Ryan Wiser said that wind, solar, and battery storage were "three legs of an important stool for near-term decarbonization" (Wiser et al., 2021).

FURTHER READING

Alonso, Elisa, Andrew Sherman, Timothy Wallington, Mark Everson, Frank Field, Richard Roth, and Randolph Kirchain. "Evaluating rare earth element availability," *Environmental Science & Technology*, Vol. 46, no. 6 (2012): 3406–3414.

American Wind Energy Association (AWEA). *U.S. Wind Industry Annual Market Reports*, April 2019.

Anchondo, Carlos. "Chevron makes historic investment in offshore wind," *E&E News*, April 14, 2021a.

Anchondo, Carlos. "BP CEO sees future in offshore wind, hydrogen," *E&E News*, May 20, 2021b.

Behr, Peter. "Can innovation save Biden's push for 100% clean grid?" *E&E News*, December 21, 2021.

Beitsch, Rebecca and Rachel Frazin. "Murkowski, Manchin introduce major energy legislation," *The Hill*, February 27, 2020.

Deveau, Scott, Sailel Kishan, and Joe Carroll. "ExxonMobil's last-ditch attempt to stave off a climate coup," *Bloomberg News*, March 29, 2021.

Earthworks. "Making clean energy clean, just & equitable," April 2021. https://www.earthworks.org/campaigns/making-clean-energy-clean/

E&E News. "Wind industry still heads for record year," April 1, 2020a.

E&E News. "U.S. solar to grow 33% this year despite COVID-19 pain," June 12, 2020b.

E&E News. "Biden's first 100 days: What's coming on energy," January 21, 2021a.

E&E News. "What Biden's $6T budget plan means for energy," June 1, 2021b.

Farah, Niina. "Groups sue to prevent birds 'getting diced' by wind turbines," *E&E News*, December 13, 2019.

Ferris, David. "EV deal shows 'Lithium Valley' could be for real," *E&E News*, July 2, 2021.

Ferris, David and David Iaconangelo. "Biden bans China imports. Will it hurt U.S. solar?" *E&E News*, June 25, 2021.

Fialka, John. "Report: U.S. solar shrugged off pandemic, enjoyed 2020 boom," *E&E News*, December 15, 2020.

Holzman, Jael and Benjamin Storrow. "Rising metals prices threaten U.S. green energy push," *Greenwire*, March 10, 2022.

Iaconangelo, David. "Do turbines hurt tourism? Depends on where you put them," *E&E News*, January 9, 2019a.

Iaconangelo, David. "Transportation crunch could threaten U.S. wind boom," *E&E News*, January 10, 2019b.

Iaconangelo, David. "Coal states prefer local wind over fossil plants," *E&E News*, March 21, 2019c.

Iaconangelo, David. "Nation's first project was good for tourism," *E&E News*, May 8, 2019d.

Iaconangelo, David. "Scientists unveil record-setting perovskite solar cell," *E&E News*, May 26, 2020a.

Iaconangelo, David. "Another Solyndra? Biden faces solar tariff test," *E&E News*, November 20, 2020b.

Iaconangelo, David. "Solar coalition: Chinese firms 'gravely' imperil U.S. clean energy," *E&E News*, August 18, 2021a.

Iaconangelo, David. "Rising costs may erase years of renewables' progress—report," *E&E News*, December 1, 2021b.

Iaconangelo, David. "Rising costs threaten U.S. solar growth—report," *Energywire*, March 10, 2022.

Iaconangelo, David and Heather Richards. "Vineyard Wind still on track despite coronavirus—BOEM," *E&E News*, April 22, 2020.

Information Technology & Innovation Foundation (ITIF). "ITIF releases report on using tax incentives to drive clean energy innovation," JEPIC, USA, December 10, 2019. https://www.jepic-usa.org/digests/2019/12/11/usaidif-releases-report-on-using-tax

Klump, Edward. "Company announces largest U.S. solar project in history," *E&E News*, November 24, 2020.

Lee, Mike. "How the oil industry is shifting to offshore wind," *E&E News*, March 12, 2021.

Marshall, Christa. "Could more wind power raise temperatures? It's complicated," *E&E News*, October 4, 2018.

Marshall, James. "Biden to order EV battery, mineral supply chain review," *E&E News*, February 24, 2021a.

Marshall, James. "Greens sue BLM to block Nev. Lithium project," *E&E News*, March 1, 2021b.

Marshall, James. "White House commits to U.S. producing clean tech metals," *E&E News*, May 27, 2021c.

Marshall, James. "It's not just mining. Refining holds U.S. back on minerals," *E&E News*, July 14, 2021d.

Mulkern, Anna. "'Record-breaking' ramp up needed for 100% clean grid—report," *E&E News*, March 16, 2021.

Richards, Heather. "Fishermen ask for shipping highways through wind farms," *E&E News*, January 7, 2020a.

Richards, Heather. "Interior: Offshore wind to have major 'adverse' effects," *E&E News*, June 10, 2020b.

Richards, Heather. "Maine plans floating offshore wind farm," *E&E News*, November 23, 2020c.

Richards, Heather. "Biden launches major push to expand offshore wind," *E&E News*, March 29, 2021a.

Richards, Heather. "Biden admin advanced R.I. offshore wind farm," *E&E News*, April 29, 2021b.

Richards, Heather. "Biden admin approves first major U.S. offshore wind farm," *E&E News*, May 11, 2021c.

Richards, Heather. "Maine picks site of 1st U.S. offshore floating wind project," *E&E News*, July 15, 2021d.

Richards, Heather and Michael Doyle. "In reversal, Interior solicitor bolsters fishermen over wind," *E&E News*, December 15, 2020.

Richards, Heather and Kristi Schwartz. "Will North Carolina go big on offshore wind?" *Energywire*, April 13, 2022.

Solar Energy Industries Association (SEIA). *U.S. Solar Market Insight: Executive Summary*, December 2019. https://www.seia.org/us-solar-market-insight

Streater, Scott. "BLM to study large-scale N.M. wind project," *E&E News*, November 12, 2018.

Swartz, Kristi. "Can solar increase emissions? A debate erupts," *E&E News*, August 21, 2019.

Swartz, Kristi. "NextEra announces historic renewable plan," *E&E News*, January 27, 2021.

Ten Brink, Talya and Tracey Dalton. "Perceptions of commercial and recreational fishers on the potential ecological impacts of the Block Island Wind Farm (US)," *Frontiers in Marine Science*, Vol. 5, no. 439 (November 2018). https://www.essoar/doi/pdf/10.1002/essoar.10500194.1

Tomich, Jeffrey. "Big solar eyes old power plants in congested Midwest grid," *E&E News*, May 5, 2021.

U.S. Department of Energy, Energy Efficiency & Renewable Energy (EERE). "Solar energy in the United States," September 2021. https://www.energy.gov/eere/solar/solar-energy-united-states

U.S. Department of Energy, Energy Information Administration (EIA). "U.S. wind energy production tax credit extended through 2021," January 28, 2022a. https://www.eia.gov/todayinenergy/index.php?tg=wind

U.S. Department of Energy, Energy Information Administration (EIA). "The United States installed more wind turbine capacity in 2020 than in any other year," April, 2020b. https://www.eia.gov/todayinenergy/index.php?tg=wind

Willson, Miranda. "Calif. rooftop solar fight could shift industry's future," *E&E News*, April 2, 2021a.

Willson, Miranda. "Lithium shortage threatens EV growth—report," *E&E News*, April 10, 2021b.

Willson, Miranda. "Want more solar panels? Good luck connecting to the grid," *Energywire*, March 16, 2022.

Wiser, Ryan, Dev Millstein, Joseph Rand, Paul Donohoo-Vallet, Patrick Gilman, and Trieu Mai. "Halfway to zero: Progress towards a carbon-free power sector," Technical report of the Berkeley Lab, U.S. DOE, April 13 2021. https://www.eurekalert.org/news-release/85654

Q13. ARE CONFLICTS OVER PUBLIC LAND USE ON THE INCREASE BETWEEN ANTI–FOSSIL FUEL GROUPS AND THE (OIL/GAS/COAL) INDUSTRY?

Answer: Yes.

The Facts: Clearly, conflicts over whether land is used exclusively, or dominantly, for oil/gas/coal exploration, development, and production have increased since the emergence of fracking technologies. These conflicts primarily concern drilling and other activities on *public*, not private lands. In the United States, some 27.4 percent of all lands are of the public domain, over which the U.S. Congress, president, and courts exercise sovereign authority. In the last 10–15 years, anti–fossil fuel protesters have indicated support for renewables, in particular solar, wind, and to a lesser extent hydropower (but rarely nuclear power). These groups have been founded in urban, suburban, and rural communities, often with a "not in my (or any) backyard" (NIMBY/NIABY) perspective. The opposition to fossil fuels may also stem from different economic or recreational interests and priorities, such as farming, ranching, or fishing, as indicated in the sample of protest activities.

Grassroots protest. Earthquakes have been linked to hydraulic fracking in several regions of Oklahoma, and they were the catalyst for organizing a grassroots organization called "Stop Fracking Payne County." Angela Spotts, a resident of Cushing (the pipeline capital of the United States), posted on her Twitter blog that there had been seven quakes in the previous hour, and called friends and neighbors to attend a meeting to do something about it. She was surprised that 105 people gathered. The group complained to an Oklahoma House of Representatives committee on earthquakes, asking for better protection of citizens and regulation of the industry. Five months later, the group's leader and followers appeared at a planning commission meeting wearing shirts declaring "They underestimate the persistence of our red dirt resistance." Ultimately, however, Spotts sold her home in Oklahoma and moved to New Mexico (McBeath, 2016; see also Dokshin, 2016).

Protests and demonstrations opposed to construction of Keystone XL and Dakota Access pipelines. TC Energy (formerly TransCanada Corporation) aspired to construct a new $8 billion pipeline from Alberta's oil sands development to Montana, South Dakota, and Nebraska. From there it would connect to other lines linked to refineries of the Gulf Coast. The Obama administration objected to the plan, but shortly after his inauguration, President Trump authorized its construction. During the 12-year controversy, both environmental and tribal groups conducted extended legal battles, protest parades, and sit-ins. Authorities commanded arrests of protesters en masse; South Dakota called protests acts of domestic terrorism. A spokesperson for the alliance of environmental and tribal groups criticized "the Trump administration's attempt to fast-track a risky tar sands export pipeline that threatens our land, water and climate" (King, 2019). One of the first acts of the Biden administration was to cancel the permit allowing the Keystone XL to cross Canada's border with the United States (Monga, 2021), shortly after which the operator shut the pipeline down. A separate pipeline protest concerned transportation of oil from the North

Dakota Bakken basin to Illinois, crossing under the Missouri and Mississippi Rivers and Lake Oahe and passing within a half-mile of the Standing Rock Sioux Reservation. Because a small section of the pipeline entered Army Corps of Engineers regulatory jurisdiction, the Sioux, other tribes, and environmental groups challenged it under NEPA, as it lacked an environmental impact statement (EIS). The protesters opposed the pipeline because of potential damage to tribal drinking water supplies and the greenhouse gas emissions from the oil it carried. The Biden administration allowed the pipeline to remain in operation while the Corps prepared a new EIS, expected to be completed in 2022 (Fortin & Friedman, 2020; Hurley, 2022).

Protests of the opening of the Arctic National Wildlife Refuge (ANWR). Under terms of the Alaska National Interest Lands Conservation Act (ANILCA), ANWR was made a wilderness area on which any development was prohibited; but 1.5 million acres of its coastal plain were designated as a study area that Congress could open for commercial development at a later time. Many Alaskans looked at ANWR as the site of the state's next oil bonanza, and for 40 years members of the Alaska congressional delegation, governors, and state legislators sought to unblock a section of the study area for oil and gas exploration, development, and production. Finally, a Republican Congress agreed on an exception to the 1980 ANILCA legislation, and President Trump happily signed it into law. Estimates on likely oil reserves varied sharply: the U.S. geological survey (USGS) said it might produce from 5.7 to 16 billion barrels; critics thought drilling would produce little oil. That ANWR lay between the oil field bonanza of Alaska's Prudhoe Bay and rich Canadian oil resources encouraged hope.

The ANWR issue divides Native communities of the North Slope and Canada. For years, the Gwich'in people living in a dozen small communities of northeastern Alaska and western Canada have opposed development. They believe it will interfere with the Porcupine caribou herd, which traverses ANWR (where it has a sacred calving area) into Canada, and which Gwich'in depend on for sustenance. The Gwich'in Steering Committee has strong ties with mainstream environmental groups (e.g., the Wilderness Society), for which ANWR has become an iconic place. However, the Inupiat Eskimos of the Alaska North Slope are shareholders in the Arctic Slope Regional Corporation (ASRC), Alaska's richest corporation, which owns subsurface rights to much of ANWR. The village corporation of Kaktovik owns surface rights, and is the only settled community in the range (Hobson, 2017). In the final days of the Trump administration, the Interior Department issued development leases for nine parcels in

ANWR at discount prices, allowing drilling to commence shortly thereafter. However, in June 2021, Biden administration Interior secretary Deb Haaland suspended the leases, asserting that they failed to comply fully with NEPA (DeMarban, 2021). As expected, most of Alaska's political leaders (including all of the three-member congressional delegation) opposed Haaland's move. (However, the Biden administration did support ConocoPhillips' proposed $6 billion Arctic project [in the National Petroleum Reserve-Alaska, NPR-A], which both environmental and some Alaska Native organizations opposed because of expansive industry operations in the reserve and apparent contradiction with Biden's "30x30" pledge [Yachnin & Richards, 2022].)

Protests of federal oil/gas regulations. As many have said, "protest is as American as apple pie," and opposition to oil/gas regulation did not start with the Trump administration. A new leasing program for Outer Continental Shelf (OCS) tracts at the end of the Obama administration brought a flood of protests to the Bureau of Ocean Energy Management (BOEM). In one case, protestors outnumbered bidders at an auction. At another, Louisiana protesters swarmed the Superdome calling loudly "Shut it down." BOEM decided to cancel public auctions and schedule them online only, citing safety concerns. In response, the Louisiana Bucket Brigade, 350.org, and other dissidents sat-in at the BOEM office with a large banner saying "MORE DRILLING = MORE FLOODING" (Yehle, 2016).

Early in the Trump administration, groups protested the direction taken by Interior Department administrators to speed commercial development of public lands (and waters). Trump's first interior secretary, Ryan Zinke, eased regulatory barriers and improved access for energy generation. He claimed federal "working lands" opened to oil, gas, and other development yielded billions in revenue and hundreds of thousands of jobs. The Center for Biological Diversity's public lands program responded by charging that "Zinke is using public lands as a cash cow to reward oil companies and other polluters, and he's locking us into a future of climate chaos" (Streater, 2018). Record-breaking lease sales in several BLM districts spurred new drilling; officials in New Mexico said they had too few staff to enforce all the rules they had. Ranchers complained that there was so much oil drilling equipment on overdeveloped land that cows had no grass to feed on. Conservationists protested that ramped-up production threatened the region's fragile and biologically diverse deserts (*E&E News*, 2018).

Biden administration actions and protest responses. The Biden's administration's aggressive (and expansive) energy agenda includes three aspects—carbon dioxide reduction, procurement of additional land for renewable energy infrastructure, and additional land for preservation of endangered and threatened species—that have been applauded by environmental groups but criticized by some industry and business groups. Interior secretary Haaland immediately formed a Climate Task Force to coordinate federal policy among governments and tribes. National Wildlife Federation CEO Collin O'Mara, meanwhile, agreed that Interior (or the Interior Department) had a "mission-critical role in our nation's transition to net-zero emissions." O'Mara emphasized that "with more than 25% of all U.S. greenhouse gas emissions originating on public lands . . . (the department has) unrivaled opportunities to restore natural carbon sinks, responsibly deploy clean energy, and reduce existing emissions" (Doyle & Streater, 2021).

Among the first actions of the Biden administration was to freeze oil/gas leasing and temporarily halt issuing permits by BLM local offices. Even as the industry howled in protest, however, oil production on public lands increased in states such as New Mexico. That was made possible by the fact that the oil industry customarily leases more public lands than it uses, essentially banking them for future development as business conditions such as rising oil prices warrant. Jeremy Nichols, the climate director for the environmental group WildEarth Guardians, said that this increased oil production "underscores that to the extent the oil and gas industry needs to drill public lands, it's certainly not feeling any pain under the Biden administration" (Richards, 2021b). Also, in the first year of the Biden administration, environmental reviews proceeded more expeditiously (Brugger, 2022).

To reach the goal of a zero-carbon economy by 2050, vast space is going to be needed to deploy wind and solar energy generation on a massive scale. The Biden administration was assisted by a Princeton University study that analyzed ways to efficiently identify 3 million megawatts of new renewable power production (nearly triple current capacity). Titled "Net-Zero America: Potential Pathways, Infrastructure and Impacts," the report contained challenging tasks, such as convincing farmers to coexist with giant turbines (Behr & Tomich, 2021). However, in some areas ideal for construction of solar panels (such as the Mojave Desert), rare and precious species with varied blooming cycles might be jeopardized by development (Yurk, 2021).

The Biden administration also proposed setting aside 30 percent of the nation's lands and waters by 2030 in an "America the Beautiful" campaign focused on protecting endangered species and conserving natural areas. Quickly labeled "30x30," the plan would require extensive coordination of the Interior and Agriculture departments. Exactly what lands would qualify for protection under such a plan immediately became a topic of considerable debate, with environmental groups expressing suspicion about the amount of input local lawmakers and commercial interests might have in its creation. Meanwhile, critics saw the 30x30 plan as hostile to oil/gas and mining interests, and ultimately as a threat to property rights (Yachnin & Cama, 2021).

The oil and gas industry protested the Biden administration's moratorium on drilling on federal land by launching media and legal campaigns. The Western Energy Alliance paid for a radio campaign in New Mexico, Colorado, Wyoming, Utah, North Dakota, and other oil/gas states that blamed the moratorium for industry job losses, higher gas prices for consumers, and less local government funding for schools and public safety. Fourteen Republican-led states challenged the Biden administration's ban in federal court; a federal district court judge agreed with the states and allowed leasing to continue nationwide (Farah, 2021). The Biden administration's "high level blueprint" for reform of the federal oil and gas program did not address climate change concerns. It did recommend a large increase in royalty rates and provisions to cover cleanup costs (Richards, 2021a). (On April 15, the Biden administration announced a resumption of oil and gas leasing on public lands in nine states [Johnson, 2022]. This broke a pledge Biden had made to climate activists to assure them that he would work to reduce use of fossil fuels, but the White House probably considered it essential as the president sought to increase U.S. oil supply after the Russian invasion of Ukraine.)

While fossil fuel advocates objected to Biden administration actions in court, climate change advocates (including celebrities such as Jane Fonda) protested in northern Minnesota streets against a controversial pipeline project (Line 3) being built by Enbridge, a leading pipeline construction company headquartered in Calgary, Alberta. The original oil pipeline, built in the 1960s, stretched from Canada's Alberta province to Superior, Wisconsin, some 1,100 miles. Line 3, the 370-mile replacement for worn-out sections in North Dakota, Minnesota, and Wisconsin, was substantially completed by September 2021. Again, the Biden administration was placed in the middle of controversy. Environmental and tribal challengers asked the Army Corps of Engineers to suspend the project until a complete environmental impact statement (EIS) was completed. The states, the oil/gas industry and its unions, and local business groups all wanted the project to proceed (Northey & Anchondo, 2021).

Traditional compromise as a resolution to land-use conflicts. Competition for scarce wilderness land does not need to lead to conflict. The 2020 Great American Outdoors Act, for example, established a National Parks and Public Land Legacy Fund to support deferred maintenance projects on federal lands. Some 50 percent of federal revenues from oil/gas/coal and alternate energy projects on federal lands/waters would be used to support the fund.

This legislation was an omnibus bill that passed both houses with comfortable (and bipartisan) majorities. It added 1.3 million acres of new wilderness, more than 600 miles of wild and scenic rivers and nearly 2,000 miles of new trails to the national system of protected areas. It also reauthorized the Land and Water Conservation Fund, which makes mandatory payments annually for conservation activity such as free entry for children and their families to parks and employment for youth working on public lands. The act gave special protections to desert wilderness; it set up new parks in the National Park Service; and it required new studies to identify special resource areas. The largest beneficiaries of the legislation are the Western states, because they have the largest percentage of federal public lands in the United States, yet Eastern states benefit as well: more than a third of the wild and scenic rivers are in Eastern states, and special landmarks and museums are designated. Altogether, every region gained something, and passage left no partisan rancor (Hotakainen, 2019).

Such legislative compromises are difficult to accomplish, however, when the leading political parties are highly polarized, as they are in the United States.

The Trump administration dismantled the Landscape Conservation Cooperative program composed of 22 regional partnerships (and administered by the Fish & Wildlife Service), and in the Biden administration resurrection of this approach is seen as useful to the 30x30 proposal. Critics from different ideological perspectives agree that a different approach emphasizing integration is preferable. A former Interior Department deputy secretary, Lynn Scarlett, speaks in favor of "a durable national conservation framework . . . that leverages the strengths of many organizations and communities and . . . advances diversity, equity and inclusion" (Yachnin, 2021). This kind of compromise will be challenging indeed.

FURTHER READING

Behr, Peter and Jeffrey Tomich. "Biden's dilemma: Land for renewables," *E&E News*, March 24, 2021.

Brugger, Kelsey. "NEPA reviews moving faster under Biden," *Greenwire*, February 14, 2022.

DeMarban, Alex. "Biden suspends ANWR oil, gas leases," *Anchorage Daily News*, June 1, 2021.

Dokshin, Fedor. "Whose backyard and what's at Issue? Spatial and ideological dynamics of local opposition to fracking in New York State, 2010 to 2013," *American Sociological Review*, Vol. 8, no. 5 (September 2, 2016): 921–48.

Doyle, Michael and Scott Streater. "Haaland revokes Trump-era orders, creates climate task force," *E&E News*, April 16, 2021.

E&E News. "Drilling overwhelms agency protecting public lands," November 14, 2018.

Farah, Niina. "14 states challenge Biden oil leasing plan," *E&E News*, March 25, 2021.

Fortin, Jacey and Lisa Friedman. "Dakota Access Pipeline to Shut Down Pending Review, Federal Judge Rules," *New York Times*, July 6, 2020.

Hobson, Margaret. "Alaska Native communities clash over ANWR bill," *E&E News*, December 4, 2017.

Hotakainen, Rob. "Sweeping bill creates many winners: 'This is a big deal,'" *E&E News*, February 15, 2019.

Hurley, Lawrence. "Dakota Access pipeline suffers U.S. Supreme Court setback," *Reuters*. February 22, 2022.

Johnson, Katanga. "U.S. to resume oil, gas drilling on public land despite Biden campaign pledge," *Reuters*, April 15, 2022.

King, Pamela. "Judges: Keystone XL construction can begin," *E&E News*, June 7, 2019.

Klump, Edward. "Sweeping bill creates many winners: 'This is a big deal,'" *E&E News*, February 15, 2019.

McBeath, Jerry A. *Big Oil in the United States: Industry Influence on Institutions, Policy, and Politics.* Boulder, CO: Praeger, 2016.

Monga, Vipal. "What is the Keystone XL Pipeline and why did President Biden issue an executive order to block it?" *Wall Street Journal*, January 21, 2021.

Northey, Hannah and Carlos Anchondo. "Line 3 is about to come online. What will Biden do?" *E&E News*, August 24, 2021.

Richards, Heather. "3 issues to watch with Biden's oil and gas overhaul," *E&E News*, November 29, 2021a.

Richards, Heather. "Biden oil well approvals outpace most of Trump era—report," *E&E News*, December 6, 2021b.

Seltzer, Molly. "Big but affordable effort needed for America to reach net-zero emissions by 2050, Princeton study shows." https://www.princeton.edu/news/2020/12/15/big-affordable-effort-needed-america-reach-net-zero-emissions-2050-princeton-study

Streater, Scott. "Interior touts economic impacts of 'energy dominance,'" *E&E News*, November 14, 2018.

Yachnin, Jennifer. "Revival planned for conservation network dismantled by Trump," *E&E News*, May 12, 2021.

Yachnin, Jennifer and Heather Richards. "Greens: Arctic oil project undermines Biden conservation goals," *Greenwire*, March 23, 2022.

Yachnin, Jennifer and Timothy Cama. "Could Sagebrush Rebellion cinders spark fire over 30x30?" *E&E News*, July 12, 2021.

Yehle, Emily. "Protesters block BOEM's New Orleans office," *E&E News*, August 23, 2016.

Yurk, Valeria. "Solar vs. species: Study warns of desert clash," *E&E News*, June 2, 2021.

Q14. ARE STATE BALLOT PROPOSITIONS TO LIMIT FOSSIL FUELS AND RENEWABLES BECOMING MORE POPULAR?

Answer: Yes.

The Facts: At the federal level in the United States, offices are filled by elections, and no votes are held on issues. At the subnational level, however, both people and ideas may be on the ballot. Twenty-four of the 50 U.S. states give citizens the right to make and/or change the law, a system called *direct democracy*. Most of these states are in the West, and they were the last territories to join the union. They entered statehood as Progressive reforms of the initiative, referendum, and recall became a popular response to concentration of capital and post-Civil War corruption. Their young constitutional cultures accommodated reform better than the Eastern, Midwestern, and Southern states, and often their elections are barometers on divisive issues.

The shale revolution gained traction and aroused opposition in the early 2000s, and since then a number of state elections have featured ballot measures related to fracking (or more broadly to fossil fuels), clean energy, and related issues (UCLA-Luskin Center, 2019). In the 2018 midterm elections, for example, more than 60 citizen-driven initiatives appeared on state ballots; in the West, a number of proposals sought to reduce reliance on fossil fuels. Most of these measures failed to pass, due in part to strong opposition from business and industry interests. Another wave of state-level ballot measures related to energy policy arrived in 2020. As in 2018, the majority of these proposals were defeated at the ballot box.

Observers contend that in many of these instances, campaign contributions from businesses and industries that support fossil fuels and oppose investments in renewables were key to defeating energy reform measures. In Colorado, for instance, opponents of a 2018 ballot proposal to subject oil and gas operations to additional zoning restrictions around schools, playgrounds, homes, and potable water sources raised $31 million, while supporters collected less than $2 million. The proposal went down to defeat.

Local ballot initiatives. A number of local campaigns have also been launched with the goal of making renewables as accessible as fossil fuels. In Columbus, Ohio, the city passed a measure to purchase renewable energy in bulk from an alternate supplier. Propositions approved by voters in Austin, Texas, included purchase of all-electric bus and bicycle fleets and investment in light rail lines. Residents of Long Beach, California, approved a measure raising the tax on oil production to improve air and water quality and to address pandemic health issues for people of color (Ge et al., 2020). Berkeley, California, voters in 2019 made hookups for natural gas illegal in new buildings. This measure inspired copycat ordinances in other cities in California and Washington state (Iaconangelo, 2021a). Finally, lawmakers in New York City approved a ban on the use of fossil fuels for building heat (for water, cooking, and building spaces), making the city the largest in the United States to pass such a ban (Iaconangelo, 2021b).

Columbia University's Climate Change Center issued a 2021 report noting that more than 100 cities, counties, and states had acted against renewable energy projects. Examples included height limits on wind turbines, bans on solar and wind farms, and stringent setback requirements, making renewable energy projects very difficult to develop (Willson, 2021). These seem clear reactions to the Biden administration's unveiling of clean energy policies and proposals. In a number of states, legislatures have used hostility to efforts to combat climate change and shifts toward green energy to actually restrict support for and development of renewables, as for example the Ohio State Senate's attempted limitation on local governments' approval of utility-scale solar and wind projects (Wagoner, 2021).

Conclusion. Land-use conflicts over fossil fuel exploration and development have increased, and far greater public attention is now centered on construction of pipelines, opening once-restricted areas (such as ANWR) to oil/gas drilling, and loosening (and in some cases eliminating) parts of the regulatory framework protecting human safety and ecological systems from contamination. In addition, the Biden administration's aggressive climate change agenda includes a major initiative to increase conservation of lands and protection of endangered/threatened species (30x30). Meanwhile, advocates of renewable energy sources have been heartened by the increase in the number of ballot propositions that have emerged in various parts of the country to support clean energy and reduce fossil fuel consumption. Some of these efforts have even been successful, despite the financial disadvantage that they often face against industry opponents. At the state and local levels, propositions on renewables have fared best in densely populated, liberal regions of the country and deep blue states such as California, Oregon, and New York. (For example, New York voters approved a ballot measure in November 2021 to add a right to a clean environment to the state constitution [Associated Press, 2021].) Ballot proposals to boost renewables have fared less successfully in other regions of the country. An annual survey by the American Council for an Energy-Efficient Economy (ACEEE) finds a large political split between red and blue states on continued use of fossil fuels vs. speedy implementation of clean energy (Behr, 2022). The causes of this important divide are still being investigated.

FURTHER READING

Associated Press. "NY voters approve right to a clean environment," *E&E News*, November 3, 2021.

Ballotpedia. "2018 ballot measures." https://ballotpedia.org/2018_ballot_measures

Behr, Peter. "Report shows red-blue state divide on clean energy," *Energywire*, February 3, 2022.

Ge, Jubing, Priyanka Roche, and Alexander Dane. "Election 2020: More state and local ballot measures advance the clean energy transition," *Red Green and Blue*, November 19, 2020. http://redgreenandblue.org/2020/11/19/election-2020-state-local-ballot-measures-advance-clean-energy-transition/

Iaconangelo, David. "Court ruling upholding gas ban transforms U.S. fight," *E&E News*, July 8, 2021a.

Iaconangelo, David. "NYC passes nation's largest gas ban," *E&E News*, December 16, 2021b.

Kutz, Jessica. "Citizens put renewable energy on this year's ballots," *High Country News*, September 27, 2018.

Lee, Mike. "Oil clout pits ethics against money in politics," *E&E News*, November 26, 2018.

Opal, Tara and Stephanie Malin. "Don't frack so close to me: Colorado voters will weigh in on drilling distances from homes and schools," *The Conversation*, September 26, 2018.

3

Alternatives

The previous chapter discussed the major renewable energy sources, two of which (hydro- and nuclear power) have declined in popularity among end users. Still, in 2020, nuclear and hydropower constituted 27 percent of total energy resources in the United States.

Two other energy sources—solar and wind—produced 10.7 percent of the electricity used in the United States in 2020. These two green options, however, have experienced the highest growth rates of energy sources in recent years with the exception of natural gas. As noted, they are bona fide renewables that do not emit greenhouse gases into the atmosphere. As a result, they draw attention and interest for both their environmental benefits and their sustainability.

There are other renewable energy sources in addition to wind and solar, however, and together they constitute about 5 percent of total U.S. energy resources. Wood products (such as cordwood, pellets, wood chips) have been used by humans for centuries for both heat and cooking, and they remain in use today. The modern term for wood products is "biomass," meaning that it is solid, not liquid, and is derived from biological organisms. Other energy sources are liquid and have started a process of decomposition or molecular change. This category is called "biofuels" and includes well-known substances such as ethanol. Both biofuels and biomass products are results of actions of the sun on the earth over the millennia and currently. Unlike fossil fuels, however, biomass and biofuel products are renewables.

The last set of renewables refers to two different processes. First is the movement of waters heated in the earth's core toward the surface, where they can be a source of both "geothermal" heat and energy. The second refers to the gravitational pull of the moon, creating strong waves twice daily: the tides themselves and waves of the ocean.

Although these various energy sources are a potpourri, they share an orientation toward the future. Even if their effect on energy supplies is slight now, observers expect them to be the subject of considerably more research and attention in the coming decades.

Four questions concerning these alternative energy resources are discussed in this chapter. The first, Q15, concerns biomass, and asks whether use of cordwood, forest, and urban wastes, will significantly increase in the next generation. Wood burned for heating and cooking composes about 15 percent of total energy resources in emergent economies, yet with high population states such as China and India becoming more economically developed, they are likely to use less biomass. But if so, will the replacement energy sources be fossil fuels?

Q16 turns to ethanol production, distribution, and consumption in the United States. Corn-based ethanol has been a bonanza for Midwestern farmers, and the biofuel has pushed corn prices higher than historic levels. This discussion starts with a brief history in the evolution of ethanol and federal policy responses to the technology. A key issue for the future of ethanol is whether a higher percentage of ethanol in fuel (E15) should be available at the pump. The advantages and disadvantages of ethanol are assessed in the context of concern about climate warming, as well as competition between the ethanol industry and the petroleum industry.

Q17 takes a different slant as it considers competitors to ethanol in the renewables sector. This area invites speculative discussion as energy resources with very low production values are being compared and contrasted. Yet resources such as hydrogen and fuel cells have stimulated scientific and technical curiosity, and increasingly are mentioned in political discourse of the Biden administration. Q18 concludes the section on renewables. It considers the science of the earth as a system and examines geothermal heat and energy generation as well as tidal/wave power.

Q15. WILL USE OF BIOMASS TO GENERATE ENERGY SIGNIFICANTLY INCREASE BY 2045?

Answer: Not likely.

The Facts: There is little likelihood that use of biomass as an energy resource will increase in the United States. Biomass is an energy source created from organic materials; its energy derives from the sun. Biomass includes a number of wood products, for example, firewood, scrap lumber, wood particles (pellets, chips), and waste wood such as forest debris. Included as well are different kinds of plant species—ryegrasses, bamboo, switchgrass—and also animal waste products (e.g., manure). The advantages of biomass use are numerous, but they have also been criticized for negative environmental effects such as pollution and land degradation.

Status of biomass in 2022. Globally, biomass constitutes approximately 10 percent of primary energy consumption, although estimates vary. An exact percentage is impossible to ascertain because many "traditional" sources—for example, forest debris, animal dung—are not measured and reported to any agency. Greatest use of biomass is concentrated in the poorest countries, because this heating and cooking resource historically has been abundant and cheap. In the United States, however, as in most other economically developed countries, only a small percentage of energy consumption—just about 2.3 percent of the total—comes from wood and wood waste (and another 2 percent from biofuels, such as ethanol, as discussed in Q16) (EIA, 2021a).

Energy Information Administration (EIA) data indicate that the greatest usage for wood and wood waste products in the United States is industrial (50 percent) and transportation (28 percent), followed by residential (10 percent), electricity generation (9 percent), and commercial (3 percent) (EIA, 2021b). Lumber and paper mills use their waste to produce electricity and steam. Residential usage, on the other hand, is mostly burning firewood or pellet logs for heat; less often, generators are powered by combustion of logs.

Advantages of wood. Wood products are easy to grow; they are renewable and sustainable. When burned, biomass releases carbon dioxide, which is a greenhouse gas. However, trees and other wood products capture through photosynthesis about the same amount of carbon dioxide that wood releases when burned. For this reason, wood products are often called a carbon-neutral energy source. Woody biomass removed from forests may reduce wildfire risks and promote rural economic development. Analysts have even speculated that biomass has a future beyond energy production, with possible uses in the production of commodities ranging from paint to lipstick to trash bags (Heller, 2018a). An additional advantage is that burning biomass waste, such as tree branches and cuttings, reduces the amount of space required for landfills and other disposal facilities.

A number of firms have explored converting plant waste to energy, such as New Jersey-based Covanta Holding Corp. A Texas firm, Systems

International, has planned commercial power plants with an oxidation process applied to a variety of waste materials. The process inventor, Steve Clark, said: "If you have a waste problem, I can convert that to fuel" (Klump, 2020). Products and refuse are easily transportable with little risk of explosion; however, they do require space for storage.

Finally, researchers have created a wood-based material that can cover exteriors of buildings, helping cool the interior without using additional energy. The material is very sturdy, white in color, and able to deflect radiation from the sun, scattering the light without creating a glare (Iaconangelo, 2019).

In the Trump administration, the departments of Agriculture, Interior, and Energy all promoted biomass as an energy resource. Trump's Energy Department estimated that from 2030 to 2040, the forestry, agriculture, and waste sectors could produce (sustainably) a billion tons of biomass annually. A meeting of the agencies' top administrators in late 2018 reported agreement on establishing "clear and simple policies to reflect the carbon neutrality of forest bioenergy" (Heller, 2018b). However, environmental groups and scientists objected to the characterization of firewood and waste burning as "carbon neutral," pointing out that such activities polluted air, land, and water. As one critic of the administration's biomass push commented, "using wood produces two to three times as much carbon per kilowatt hour as burning coal or natural gas," and it takes decades of forest regrowth to offset carbon released in burning (Searchinger, 2020).

Disadvantages. Problems in combustion of biomass have appeared greater than advantages in the last several decades. When burned, these commodities release chemicals and particulates into the air, some of which are hazardous to human health and to ecosystem sustainability. Wood smoke contains carbon monoxide and small particles that lodge easily in human and animal lungs. Wood smoke particulates can be especially dangerous to individuals with respiratory difficulties such as asthma, allergies, and shortness of breath. Also, it increases risks of heart failure for those with cardiovascular difficulties and increases propensity to mental depression and dementia (see Q31).

Use of scrubbers in boilers can clean emissions, and cloth filters can remove particles from smoke. In addition, burning at very high temperatures breaks down chemicals into less dangerous compounds, and public health regulations also can come into force to reduce the environmental impact. However, producing an entirely clean emission from wood burning is not yet realistic.

A second environmental disadvantage to biomass is the impact that demand has on existing stands of forests. In other words, extensive biomass consumption can end up encouraging clear-cutting forests to harvest timber. This form of harvesting is most economically efficient, but it disrupts and degrades ecosystems, wildlife habitat, and numerous species of flora and fauna and can make wooded areas more vulnerable to wildfires. Another particularly damaging impact of clear-cutting is soil erosion. Rivers, streams, and lakes can all suffer when soil that was once held in place by tree root systems pours into life-sustaining waterways. A third disadvantage is that when land is used for tree plantations, it is unavailable for potentially more beneficial uses, such as raising food crops.

Deregulation of wood stoves. The Trump administration pursued a deregulatory agenda more aggressively than had previous Republican and Democratic administrations on a wide range of fronts. In late 2018, for example, the White House Office of Information and Regulatory Affairs (OIRA) proposed a new EPA rule that would grant a partial reprieve of tough emissions standards for new wood-fired home heating systems, notwithstanding lack of evidence that benefits (for the wood products industry) would exceed costs (of greater pollution and health losses). The Trump administration's EPA approved the waiver, but it was quickly challenged by environmental groups in the courts (Reilly, 2018).

Meanwhile, as part of a Trump administration policy to fast-track requests for regulatory relief because of the COVID-19 pandemic, the EPA proposed a draft rule to temporarily permit sale of stove models not meeting strict emissions limits. Senator Tom Carper (D-Del), the ranking member of the Senate's Committee on Environment and Public Works, asked EPA administrator Andrew Wheeler for clarification on reports that the agency was considering using the coronavirus crisis as a pretext to allow retailers to sell wood heating systems that failed to meet clean air standards. Rising to defense of the agency was the stove manufacturers' trade group, the Hearth, Patio and Barbecue Association. One association spokesperson insisted that the noncompliant models were "exceptionally clean-burning" and could replace much dirtier-burning appliances (Reilly, 2020a).

At EPA's hearing on the proposal, Maria Smilde of the environmental group Earthjustice declared that "the EPA should not use one public health crisis to rationalize extending another" (Reilly, 2020b). The agency then confused the issue of the safety of burning biomass when in a submission to the court it noted that "Biomass co-firing does not reduce CO_2 emissions at the source—it increases them." This was in seeming contradiction to the agency's position that biomass was carbon neutral (Heller, 2020c). The tougher standards took effect in May 2020 as scheduled, but the Trump administration said it would do little to stop the sale of higher-polluting models that dealers had in stock.

Biomass policies of the Biden administration. After Joe Biden was sworn in as America's 46th president in January 2021, he moved quickly to rescind as many of the former president's last-minute regulatory actions as possible. This included Trump's decision to leave national particulate standards unchanged.

In early 2021, the EPA announced approval of a new test method for wood stoves (EPA, 2021a) and an April 2021 advisory that "the agency is taking a number of actions to address concerns about the methods and manner in which new wood stoves are being tested" (EPA, 2021b). In August 2021, a three-judge panel for the U.S. Court of Appeals for the District of Columbia ruled in favor of the agency.

The EPA's action linked the inadequacy of Trump administration rules to requirements of the Clean Air Act (see Q31), and research connecting exposure to fine particulates and increased vulnerability to COVID-19. Biden's EPA administrator, Michael Regan, also referred to environmental justice objectives in defending the administration's firmer stand on particulate pollution: "The most vulnerable among us are most at risk from exposure to particulate matter, and that's why it's so important we take a hard look at these standards that haven't been updated in nine years" (Reilly, 2021a). Among Regan's first actions as administrator was dismissing members of the Clean Air Scientific Advisory Committee (CASAC)—a group organized by Trump administration officials that included members with obvious conflicts of interest to the regulated industry.

The EPA said its system assured that wood-fired appliances performed at the level required by federal standards. However, a group of state air quality regulators challenged EPA's system, saying it "provides no confidence" stoves meet required standards (Reilly, 2021b). In March 2022, the agency's Clean Air Scientific Advisory Committee (CASAC) concluded its review of national soot standards. All seven CASAC members agreed with career staff that evidence warranted tightening of the daily and annual soot exposure limits (Reilly, 2022).

FURTHER READING

American Lung Association. "More than 4 in 10 Americans breathe unhealthy air, people of color 3 times as likely to live in most polluted places," Chicago, April 21, 2021. https://www.lung.org/media/press-releases/sota-2021

Heller, Marc. "The future of biomass is lipstick and trash bags," *E&E News*, October 31, 2018a.

Heller, Marc. "Agency heads team up on biomass," *E&E News*, November 1, 2018b.

Heller, Marc. "Carbon neutral claims could snag biomass rule," *E&E News*, July 27, 2020.

Iaconangelo, David. "'Cooling wood' could slash electricity use," *E&E News*, May 24, 2019.

Jacobs, Jeremy. "Court extinguishes challenge to EPA wood heater rule," *E&E News*, August 27, 2021.

Klump, Edward. "The next low-carbon energy source? It might be trash," *E&E News*, February 18, 2020.

Reilly, Sean. "White House clears EPA wood stove proposals," *E&E News*, November 19, 2018.

Reilly, Sean. "Carper questions EPA's plans to delay wood stove standards," *E&E News*, May 11, 2020a.

Reilly, Sean. "EPA packs 'economic relief' for industry into stove rule," *E&E News*, May 28, 2020b.

Reilly, Sean. "Pandemic-era wood stove relief decried as 'irresponsible,'" *E&E News*, June 8, 2020c.

Reilly, Sean. "Report finds 'systemic failure' in EPA wood stove tests," *E&E News*, March 18, 2021a.

Reilly, Sean. "EPA to rethink soot standard Trump left unchanged," *E&E News*, June 11, 2021b.

Reilly, Sean. "EPA panel backs tighter soot standards," *Greenwire*, March 22, 2022.

Ruskin, Liz and Emily Holden. "Natural but deadly: Huge gaps in US rules for wood-stove smoke exposed," *The Guardian*, March 16, 2021.

Searchinger, Tim. "Op-ed: Is burning wood for power carbon-neutral? Not a chance," *Los Angeles Times*, December 28, 2020.

U.S. Department of Energy, Energy Information Administration (EIA). "Biomass explained: Wood and wood waste," May 11, 2021a.

U.S. Department of Energy, Energy Information Administration (EIA). "Biomass explained: Biomass and the environment," June 8, 2021b.

U.S. Environmental Protection Administration (EPA). "Controlling air pollution from residential wood heaters," April 5, 2021a. https://www.epa.gov/residential-wood-heaters

U.S. Environmental Protection Administration (EPA). "EPA-approved test labs and third-party certifiers for residential wood heaters," April 16, 2021b. https://www.epa.gov/burnwise/epa-approved-test-labs-and-third-party-certifiers-residential-wood-heaters

Q16. WILL A HIGHER PERCENTAGE OF ETHANOL IN FUEL (E15) DOMINATE THE TRANSPORTATION ENERGY MARKET BY 2025?

Answer: Unlikely.

The Facts: Corn-based ethanol is a relatively new source of energy. The ethanol story is interesting because of the extensive government involvement in its production. Without continued protective regulation by the federal government, the industry would collapse. This made the industry susceptible to vicissitudes within the Trump administration, especially evident after early 2020, during which time fossil fuel prices crashed temporarily and the COVID-19 pandemic erupted across the United States and around the world. Many of the issues of that period continue into the Biden administration, but now the competitive pressure for ethanol comes from renewables and not fossil fuels.

The term ethanol is derived from "ethane," an odorless, colorless, and flammable gas. It is a type of biofuel that, when blended with gasoline or diesel fuel, provides fuel for vehicles in the transportation system. A number of organic, mostly plant-based materials, can be used to produce alcohol fuels. The sugars found in grains such as corn, barley, and sorghum are the primary source of ethanol. Other sources include sugarcane, sugar beets, potato skins, yard clippings, rice, tree bark, and switchgrass (EIA, 2021).

The corn-based ethanol industry is the largest biofuel industry in the United States. In 2014, it produced more than 14 billion gallons, and most cars in the United States had at least 10 percent ethanol in their tanks. Ethanol has greatly benefitted American corn farmers by raising agricultural commodity prices, which has brought about increased economic activity in rural areas. Corn is extensively grown in the U.S. Midwest, and these corn belt states have become champions of ethanol production and supporters of government subsidies to boost the industry.

Government assistance for ethanol production. Interest in ethanol soared after the onset of the energy crisis in the 1970s. Policymakers' strategy was to promote growth of ethanol, and they used as their tactics a series of tax credits and other economic incentives. Lawmakers argued that corn-based ethanol would reduce air pollution, and even more importantly reduce dependence on foreign oil. The first ethanol tax credit was adopted in 1978 (Duffield et al., 2015), followed by tariffs and duties on ethanol produced in other countries. To increase demand for U.S.-grown ethanol, in 1988 Congress passed the Alternative Motor Fuels Act. The act gave credits to automakers for meeting corporate average fuel efficiency (CAFE) standards if they manufactured "flexible-fueled" vehicles (FFVs) that could operate on gasoline or any combination of gas and ethanol (Duffield et al., 2008; see also Q32).

The measures led to slow, steady growth in ethanol production until the early 1990s, when new environmental legislation gave ethanol a large boost. The Clean Air Act Amendments of 1990 (CAA) required both a reformulated petroleum product (for which ethanol later became a substitute) and a reformulated gasoline program, to control pollution problems in urban areas not compliant with clean air standards. In addition, the 1993 amendment required federal agencies to purchase a percentage of alternative-fuel vehicles fueled by propane, natural gas, and ethanol (Duffield et al., 2015).

Funding bills for the Department of Agriculture (USDA) in the late 1990s further expanded biofuels. One provision authorized pilot projects for harvesting biomass on lands set aside for crop production under the Conservation Reserve Program (CRP) (Collins & Duffield, 2006). USDA also started a Commodity Credit Corporation (CCC) Bioenergy Program to stimulate ethanol demand and reduce crop surpluses. This program made cash payments to eligible ethanol producers.

Rising oil prices in the early twenty-first century rekindled congressional interest in ethanol. Corn-based ethanol benefitted greatly from the 2004 Jobs Act, which created an ethanol excise tax credit. The Jobs Act gave oil companies flexibility to blend any amount of ethanol into gasoline to meet their octane and oxygenate needs (Duffield et al., 2008). One year later, in the Energy Policy Act of 2005, Congress adopted a renewable fuel standard (RFS) with biofuel production mandates. Refiners and marketers could meet requirements of federal legislation with ethanol alone.

Then, as oil prices continued to climb, Congress passed another energy bill called the Energy Independence and Security Act (EISA) of 2007. It replaced the RFS with a more aggressive set of renewable fuel mandates called RFS2. Ethanol by then had become a conventional fuel category. To broaden ethanol production beyond corn-based ethanol and further reduce greenhouse gas emissions, Congress added a volume requirement for an advanced biofuel, called "cellulosic biofuel" (see Q17).

In 2015, the renewable fuel standard for corn ethanol was capped, while advanced biofuels mandates increased (which diversified feedstocks used to produce renewable fuels). Revised standards included a greater variety of substances: rendered fats and greases, oil from algae and biodiesel, as well as sugarcane ethanol. The U.S. standard thus has gone beyond corn-based ethanol and increased use of cellulosic biomass (EIA, 2012). In the last decade, the corn-based ethanol industry has become mature. At the outset in the 1970s, only a few small firms were in business; 40 years later there were 209 ethanol plants spread across 29 states. Most gasoline sold in America today contains about 10 percent ethanol.

EIS and the challenge to ethanol. In 2001, the EPA approved use of 15 percent ethanol (called E15), and the market for it began to grow, reaching 100 stations in 16 states by 2015 (Duffield et al., 2015). The question is whether E15 (with 15 percent ethanol and 85 percent gasoline) is likely to become dominant in the future. Early in 2019, the Renewable Fuels Association (RFA) initiated a campaign for a higher percentage of ethanol in fuel, asking the EPA to make final its approval for E15 year-round. Advocates of a final E15 rule contend that ethanol burns more cleanly than gasoline, and that it releases lower carbon emissions than petroleum. Further, supporters of increasing the ethanol quotient say that it would boost demand for corn, raising its price for the benefit of farmers and encouraging economic development in rural areas. Anticipating the argument that vehicles would need to be adapted to use E15, advocates opined that more than 90 percent of new models' engines could accommodate the fuel (Heller, 2019).

Opponents of a final E15 rule believed it was unnecessary, because E10 already was a standard blend. They did not want to see any further increase in ethanol percentage until the system of renewable fuel credits was changed. Opponents also contended that the blending requirements for ethanol caused hardship to small refineries (Heller, 2019). An additional argument against the E15 rule rested on principles of market freedom from government interference. Several senators from oil states urged the EPA to reduce the corn-based ethanol it required to be available, saying that the "blend wall" (maximum amount of biofuel feasibly mixed into the fuel supply) ran against market realities, and that consumers wanted the freedom to use no ethanol at all (Heller, 2019). Opponents also made the case that an ethanol requirement was no longer necessary because the United States had become an oil/gas-exporting nation and appeared likely to remain so.

Fossil fuel v. ethanol competition. The advocates and opponents of an expanded ethanol mandate corresponded closely to the division between the petroleum industry (represented by the American Petroleum Institute [API]) and the corn-based ethanol industry (represented by the Renewable Fuels Association [RFA]). Although the Trump administration had reservations about renewables, the biofuel issue was distinct because farm states and rural areas, where support for ethanol is strong, were among the strongest supporters of Trump in particular and Republicans in general. Aware of this political reality, Trump tweeted "The Farmers are going to be so happy when they see what we are doing for Ethanol" (Heller, 2019).

The Trump administration offered the ethanol industry a streamlined process for expanded E15, retention of mandates for ethanol, and compensatory benefits for farm states that were hurt by Trump administration trade policies with China. As part of its "America First" strategy, the Trump administration set and escalated tariffs on industrial goods imported from China. China retaliated by imposing tariffs on U.S. agricultural exports (soybeans, corn, other crops). American farmers raising these crops lost income as demand fell (corn prices dropped by about 10 percent), and farmer bankruptcies subsequently rose.

Although ethanol was important to farm states, the petroleum industry had significant clout in states like Texas and Pennsylvania with large numbers of Electoral College votes. Perhaps this factor accounted for the Trump administration's final rule, in which it only granted waivers of biofuel-blending requirements to small refineries in certain situations (Heller, 2019). This change failed to satisfy farmers, ethanol producers, or their trade associations. On the campaign trail in September 2020, candidate Joe Biden attacked Trump's record on ethanol, remarking that the moves he made to assist the industry were "too little, too late and transparently political" (Kelly, 2020).

Perhaps sensing the need to fortify their position in their competition with fossil fuels, in early 2020 ethanol producers temporarily joined forces with agricultural associations, electric vehicle advocates, gas and electric utilities, and clean energy advocates in a coalition calling for decarbonizing transportation. The coalition issued a white paper describing a "low-carbon fuel standard" for the Midwest that would reward fuels or technologies based on their carbon emission reductions (*E&E News*, 2020).

The COVID-19 pandemic and a new role for ethanol. At the outbreak of the coronavirus pandemic, hospitals, clinics, and nursing homes were desperate to locate essential medical supplies including hand sanitizers. Ethanol is derived from the conversion of cornstarch to glucose and its combination with yeast, and is a common ingredient in sanitizers. It appeared to be an ideal supplier, especially as the industry could produce it in huge quantities. Initially, however, the Food and Drug Administration (FDA) objected to its use on two grounds. FDA standards required approved products to meet high-quality specifications for use in drugs or beverages. And it objected to corn-based ethanol because the alcohol was not denatured (mixed with a bitter supplement to make it undrinkable). Such modifications were essential to avoid poisoning children who mistakenly imbibed sanitizers (*E&E News*, 2020).

The Renewable Fuels Association (RFA) joined with the Consumer Brands Association to ask the FDA to reconsider its guidelines, and the agency capitulated on the first issue. It continued to require that a bitter additive be added to the sanitizer, however. Geoff Cooper, CEO of RFA, said: "To FDA's credit, they did take to heart some of the concerns that were being raised by the industry and made some slight changes and modest tweaks to their guidelines" (*E&E News*, 2020). More than a dozen plants responded to the new FDA rules by adding ethanol to their sanitizing products.

Biden administration changes. Ten Midwest states produce the lion's share of ethanol in the United States. While Joe Biden won 306 electoral college votes (to Trump's 232) in the November 2020 presidential election, he won just three of those top ethanol-producing states for a total of 41 electoral college votes (Trump won 51 electoral votes from the other seven) (National Archives, 2020).

In the summer of 2021, Congress delivered a massive infrastructure investment bill to the Biden administration. The bill, which received bipartisan support, included measures clearly beneficial to American ethanol producers. EPA administrator Michael Regan also pledged that biofuels would be a key component of the agency's climate strategy, despite the Biden administration's push for greater investment in electric vehicles (McCrimmon & Tamborrino, 2021).

Lobbying groups for ethanol applied pressure on the Biden administration in its second year to make E15 available year-round. Growth Energy ran ads on CNN, MSNBC and Fox News making the argument that E15 was no riskier to use than 10 percent ethanol sold in gas stations throughout the year; they also pointed out that E15 sold for less per gallon than regular gasoline (Heller, 2022). However, it was soaring gas prices and their likely adverse impact on 2022 midterm congressional elections that most favored E15 supporters. The EPA granted an emergency waiver to allow E15 to be sold in the summer months. A White House official said the waiver "can give American families more flexibility and offer real savings at current prices." While this move did not win the Biden administration support from Senator Grassley (R-IA), the chief advocate of E15, the CEO of the American Coalition for Ethanol called it "great news." Climate activists, on the other hand, were indignant about what they thought would be adverse impacts on air quality and public health (Bravender, 2022).

FURTHER READING

Bravender, Robin. "Biden turns to ethanol to ease gas prices," *Greenwire*, April 12, 2022.

Collins, Keith and James Duffield. "Evolution of renewable energy policy," *Choices*, Vol. 21, no. 1 (2006): 9–14.

Duffield, James, Irene Xiarchos, and Steve Halbrook. "Ethanol policy: Past, present, and future," *South Dakota Law Review*, Vol. 53, no. 3 (2008): 425–83.

Duffield, James, Robert Johansson, and Seth Meyer, eds. *U.S. Ethanol: An Examination of Policy, Production, Use, Distribution, and Market Interactions.* Washington, DC: U.S. Department of Agriculture, 2015. https://www.usda.gov/oce/energy/index.htm

E&E News. "Ethanol and electric vehicle advocates merging on fuel plan," March 16, 2020a.

E&E News. "Ethanol plants seek rule changes to resupply hand sanitizer," March 27, 2020b.

E&E News. "FDA changes boost alcohol for sanitizer from ethanol makers," April 1, 2020c.

Heller, Marc. "Ethanol group launches campaign for final E15 rule by summer," *E&E News*, January 15, 2019a.

Heller, Marc. "Oil industry fumes as USDA chief champions ethanol," *E&E News*, April 22, 2019b.

Heller, Marc. "Senators urge EPA to limit ethanol mandate," *E&E News*, May 1, 2019c.

Heller, Marc. "Trump promises 'giant package' for ethanol industry," *E&E News*, August 28, 2019d.

Heller, Marc. "EPA releases final rule that angers ethanol, farm boosters," *E&E News*, December 19, 2019e.

Heller, Marc. "Ad campaign urges expanded E15 to ease high gas prices," *Greenwire*, March 25, 2022.

Kelly, Stephanie. "U.S. presidential candidate Biden rips Trump's record on ethanol," *Reuters*, September 15, 2020.

McCrimmon, Ryan and Kelsey Tamborrino. "Midwest farmers look to plow through Biden's electric-vehicle push," *Politico*, May 4, 2021.

National Research Council. *Renewable Fuel; Standard: Potential Economic and Environmental Effects of U.S. Biofuel Policy.* Washington, DC: National Academies Press, 2011.

OUP. *The New Shorter Oxford English Dictionary*. New York: Oxford University Press, 1993.

U.S. Energy Department, Energy Information Administration (EIA). *Biofuels Issues and Trends*. October 2012. https://www.eia.gov/biofuels /issuesstrends/pdf/bit.pdf

Q17. ARE ALTERNATIVE FUELS AND GASES FROM RENEWABLE SOURCES LIKELY TO BECOME MORE IMPORTANT FOR HEATING/ ELECTRICITY GENERATION BY 2045?

Answer: Yes.

The Facts: The remaining alternative plant- and animal-based energy sources contribute less than 1 percent to the U.S. national energy profile, but they are the subject of considerable research, and their contribution to America's energy portfolio seems highly likely to increase. Six examples present diverse sources: biodiesel, biogases (from biomass), methane-to-methanol conversion, cellulosic ethanol, the gas/liquid hydrogen and fuel cell technology, and nuclear fission.

Biodiesel. This energy source is made from animal fat or plant oils (such as soybean oil, canola, or waste oil from food production). It can be used in its pure form, as is the practice in Europe. In the United States, biodiesel is typically a fuel additive. Since the 2005 energy legislation, biodiesel has been eligible for a tax credit (part of the renewable fuel standard [RFS]), without which producers say they operate at a loss. Congress usually allows the credits to expire and then extends them retroactively, an arrangement that forces biodiesel companies to operate and make business decisions without knowing if the credits will continue (Heller, 2019).

The attractiveness of an increase in biodiesel fuel depends on the feedstock, because large-scale production of some potential biodiesel fuels (such as soybeans) may increase greenhouse gas (GHGs) emissions. In addition, using feedstock plants that are inefficient in their processing of fertilizers and water may result in increased land and water pollution. These are some of the reasons why most forms of biodiesel will continue to be used as additives to diesel fuel, as opposed to standing alone as a fuel alternative in its own right (Reisser & Reisser, 2019).

Methane conversion to methanol. Methane is the main component of natural gas. In the course of oil and gas production or during transport, methane becomes a heat-trapping gas. Methanol is produced by slightly altering the chemical composition of methane. Methanol is a liquid fuel that can be used to produce power for vehicles such as trucks and buses, or it can become a feedstock used to make other chemical products. Several species of bacteria ingest methane, producing methanol, and researchers have conducted a variety of tests to explore ways to convert methane into fuels and chemicals (Mandel, 2019).

Facilities used to convert methane to methanol have a large footprint and cannot be set up easily at small oil and gas drill pads where gas is now flared, releasing GHGs into the air. This was the situation encountered by Northwest Innovation Works, which proposed a $2.3 billion project to take fracked natural gas from Canada and convert it into methanol. The Chinese government backed this seven-year plan to build one of the world's largest methanol plants along the Columbia River in Washington state. The methanol would be shipped to China as an ingredient for plastics to make products such as iPhones, apparel, and medical devices. Although the project was initially supported by Washington governor Jay Inslee, regulatory officials paid more attention to environmental groups who warned that "the climate-killing emissions from this project would have overwhelmed Washington" (*E&E News*, 2021).

Biogas. Another type of biofuel is biogas, a combustible fuel produced from relatively recent (not fossil) living matter. It is primarily a mixture of methane, carbon dioxide, and other trace gases that form from degradation of organic matter in environments where little oxygen is present. Examples of sources include cores of some trees, water sediments, decayed animals and insects, landfills, wastewater storage facilities, and crops such as sugar cane and peanuts (IEA, 2020)

Biogases are utilized as transportation fuels, primarily in trucks and buses. They are also used in fuel cell generators and as direct sources of heat. Compared to corn- and sugarcane-based ethanol, biogas is more efficient in capturing energy. Biogas plants are not necessarily good neighbors, though. The Big Ox biogas plant in Nebraska processes organic waste from food and beverage manufacturers into methane. It also treats and discharges wastewater from other industries. Neighbors complained about odors from the factory, ownership accrued unpaid fines and penalties of nearly $1 million, and the state eventually revoked its air and storm water permits, forcing it to close (*E&E News*, 2020).

The methane content of biogas ranges from 45 to 75 percent by volume, with the remainder being carbon dioxide mostly. Biomethane (also known as "renewable natural gas") is a source of methane usually produced by upgrading biogas (removing carbon dioxide and other contaminants). Recently, biomethane has become newsworthy because oil multinationals have bought dairies and converted cow manure into renewable natural gas (RNG) as a way to reduce their carbon footprint (while stinking up the neighborhood). For example, Chevron and a joint venture partner, Brightmark (a waste solutions manager), planned to build 10 facilities to produce dairy biomethane to fuel long-haul trucks. Another oil multinational, Royal Dutch Shell, had similar plans. While the oil companies assert that they are reducing carbon in transportation fuels, critics say this has little to do with action against climate warming and instead is an excuse to keep fossil fuels in the energy mix (Bloomberg, 2021).

Cellulosic biofuel. In the Energy Independence and Security Act of 2007, Congress sought to broaden ethanol production beyond corn-based ethanol. To that end, it added a requirement for production of cellulosic biofuel, a renewable fuel derived from cells and fibers of plant tissues. To receive government subsidies, this biofuel had to meet a 60 percent reduction threshold in greenhouse gases, which distinguishes it from corn-based ethanol and biodiesel. Sources of the cellulosic feedstocks include residues of agricultural products—for example, forestry biomass, urban waste, switchgrass, and fast-growing trees such as poplar (Duffield et al., 2015).

Hydrogen and fuel cells. Hydrogen is an intriguing element in America's energy picture. It has applications for electricity generation in fuel cells and vehicles, and studies have estimated that 1 billion metric tons of hydrogen could be produced annually from solar, wind, and biomass resources. The cost of these fuel cells is high, however, presenting a major barrier to widespread adoption. Because the amount of fuel cells produced to the present is small, there are no economies of scale that would reduce prices. However, the consultant firm Deloitte, in combination with a fuel cell producer, published a white paper forecasting that by 2026, lifetime costs of fuel cells for heavy-duty vehicles (forklifts, buses, vans, long-haul trucks) would drop below those of electric vehicles—and that by 2027, lifetime costs of these fuel cells could fall below the lifetime costs of fossil fuels for heavy vehicles (Iaconangelo, 2020).

High fuel cell prices are not the only challenges, however. The infrastructure needed for hydrogen to displace gasoline consumption is not available and is unlikely to be built where demand is strongest. In 2022, there are just a few hydrogen filling stations, and hydrogen-powered vehicles cannot be used beyond these locations. Nor are there pipelines to move hydrogen cross-country. To make hydrogen competitive with fossil fuels would require building an extensive network of service stations and pipelines. This raises the issue of risk, because hydrogen is a rocket fuel. It ignites more easily than natural gas and has an invisible flame. When transported via pipelines, it can even make the pipeline's steel material brittle (Soraghan, 2021).

In Europe and other regions, alternative energy sources can compete with fossil fuels because the state assesses carbon taxes on oil and other fossil fuels, but the United States is not yet fertile ground for this type of regulatory tax. However, the Biden administration has now joined 42 other countries in supporting a series of specific technologies such as "clean" hydrogen and renewable power in different places and climates.

Interesting research programs related to hydrogen's energy potential have also been announced in recent years. In 2020, for example, industrial gas company Air Products and Chemicals Inc. announced its partnership with a Saudi Arabian firm to construct the "world's largest green hydrogen" project for the futuristic planned city of Neom (Willson, 2020). But large technological challenges still need to be overcome to make hydrogen a more viable alternative, such as carbon storage when hydrogen is produced from natural gas (*E&E News*, 2020).

Notwithstanding the economic feasibility problem, hydrogen remains attractive because there are ways to produce it that are "emissions friendly," both as a transportation and electricity generation fuel. Provided they include effective means to capture and store carbon so as to reduce emissions of the greenhouse gas carbon dioxide, such production methods could create "blue hydrogen" that would entail far fewer pollution risks to land, air, and water. A critical review of existing operations, though, argues that greenhouse gas emissions from production of blue hydrogen are very high, especially due to the release of "fugitive methane"—leaks of methane into the atmosphere (Pannett, 2021). The response to this criticism was that for the near term, blue hydrogen would be the best option, notwithstanding some greenhouse gas emissions, because its cost would be cheaper than hydrogen derived solely from wind and solar for most of the 2020s. It would be a "bridge" renewable that could be used until even cleaner energy sources emerge (Iaconangelo, 2021a).

Jeremy Rifkin believes that nations will never attain sustainability in the fossil fuel era, and argues instead for a hydrogen economy built over a "decentralized and democratized energy web" (Rifkin, 2002). In fact, the Department of Energy's "Earthshots" initiative aims to make low-carbon hydrogen cheaper than the "dirty" kind made from natural gas by 2030. When Biden administration energy secretary Granholm issued a call for "viable hydrogen demonstrations," she was indeed responding to Rifkin's plea (Iaconangelo, 2021b). The twists and turns of hydrogen policy evolution meant that hydrogen was being taken seriously as an alternate energy source.

In a 2022 report, the International Renewable Energy Agency (IRENA) predicted a "disruption" of energy geopolitics in the coming decades as hydrogen emerged as a serious fuel for climate policy. IRENA's director general, Francesco LaCamera, commented it was a game changer to achieve carbon neutrality. "But hydrogen is not the new oil." Its disadvantages included a huge increase in demand for electricity, a possibility to add to water use conflicts, and technologies that do not yet exist at scale (e.g., gas turbines capable of burning pure hydrogen). Yet the advantages of hydrogen more than compensate for these defects. It could account for 20 percent of reductions needed for a global net-zero emissions system by 2020. Unlike fossil fuels, trade in hydrogen is "unlikely to become weaponized or cartelized." And it may play a significant role in manufacturing of electrolyzers, which might create more than a million jobs in the next decade (Iaconangelo, 2022).

Nuclear fusion. Q11 considered nuclear fission, the splitting of nuclei of atoms into two fragments (of approximately equal mass, and its conversion into energy). Nuclear fusion, on the other hand, occurs when two atoms bang together to form a heavier atom. This process creates huge amounts of energy, but does not produce radioactive particles (EIA, 2021). The fusion process involves collisions of isotopes of hydrogen and a superheated gas called a plasma.

Scientists have studied fusion reactions, but have found them to be difficult to sustain for lengthy periods because of the enormous amounts of pressure and exceedingly high temperatures required to join the nuclei together. Nonetheless, the National Academies of Sciences, Engineering, and Medicine has called on the Nuclear Regulatory Commission to develop a regulatory plan for a fusion power plant by 2027, and its completion by 2040.

Motivating fusion research is protection of the U.S. power grid from blackouts, which may increase with rising use of solar and wind energy, both of which rely on systems that do not recover from blackouts easily). The proposed fusion plant would cost \$5–6 billion; it would be able to deliver energy on demand to different sections of a region, through a process called "blackstart" (Fialka, 2021).

FURTHER READING

Bloomberg. "Chevron expands push into dairy-to-gas fuel to cut emissions," *E&E News*, August 25, 2021.

Duffield, James, Robert Johansson, and Seth Meyer, eds. *U.S. Ethanol: An Examination of Policy, Production, Use, Distribution, and Market*

Interactions. Washington, DC: Office of Energy Policy and New Uses; U.S. Department of Agriculture, September 2015. https://www.usda.gov/oce/energy/index.htm

E&E News. "Neb. revokes permits for strong-smelling biogas plant," January 24, 2020.

E&E News. "Company cancels $2.3B Wash. methanol plant," June 14, 2021.

Fialka, John. "National Academies calls for first U.S. fusion plant by 2020," *E&E News*, February 15, 2021.

Heller, Marc. "Lawmakers intensify pressures on EPA over electricity fuels," *E&E News*, November 19, 2019.

Howarth, Robert and Mark Jacobson. "How Green is blue hydrogen?" *Energy Science & Engineering*, Vol. 9, no. 10 (October 2021): 1676–87.

Iaconangelo, David. "Hydrogen trucks could best EVs on cost, emissions," *E&E News*, January 13, 2020.

Iaconangelo, David. "Is 'blue' hydrogen clean energy? Studies stir climate debate," *E&E News*, August 14, 2021a.

Iaconangelo, David. "Granholm launches 'big, hairy, audacious' hydrogen plan," *E&E News*, June 8, 2021b.

Iaconangelo, David. "'Not the new oil,' Report maps barriers for clean hydrogen," *Energywire*, February 28, 2022.

International Energy Agency (IEA). "Outlook for biogas and biomethane: Prospects for organic growth," 2020.

Mandel, Jenny. "Researchers trying to turn methane into 'sustainable' fuel," *E&E News*, May 17, 2019.

Pannett, Rachel. "It's hailed as the clean energy of the future. But hydrogen produces 'substantial' emissions, study shows," *Washington Post*, November 18, 2021.

Reisser, Wesley and Colin Reisser. *Energy Resources: From Science to Society*. New York: Oxford University Press, 2019.

Rifkin, Jeremy. *The Hydrogen Economy: The Creation of the World-Wide Energy Web and the Redistribution of Power on Earth*. New York: Putnam, 2002.

Soraghan, Mike. "Hydrogen could fuel U.S. energy transition. But is it safe?" *E&E News*, August 20, 2021.

U.S. Department of Energy, Energy Information Administration (EIA). "Fission and fusion: What is the difference?" April 1, 2021.

Willson, Miranda. "Company announces massive $5B 'clean hydrogen' plant," *E&E News*, July 8, 2020.

Q18. WILL GEOTHERMAL AND TIDAL/WAVE POWER SIGNIFICANTLY INCREASE IN THE U.S. ENERGY PROFILE BY 2045?

Answer: Yes.

The Facts: Energy experts believe that use of geothermal and tidal/wave power will increase substantially in the next generation and become more competitive with other renewables and fossil fuels. The energy sources surveyed in this chapter are responsible for the smallest share of the U.S. energy mix, less than 5 percent. They are naturally occurring and will not disappear in one or two generations. Most of the sources emit few or no greenhouse gases, and thus do not put at risk the health of the earth itself. Most do not endanger species under the protection of the Endangered Species Act, or their habitats, and most do not endanger ecosystems.

However, the technology needed to produce heat by use of these alternatives is still under development and much more costly than the technology needed to drill oil wells in the Permian Basin. Moving these resources to market will require tremendous investments in transportation infrastructure, and the energy products themselves will require dispensing stations. Both economic and technological challenges will limit the speed in development of these alternatives.

Geothermal energy. The genesis of geothermal energy is slow decay of radioactive particles deep in the earth's inner core. Some 4,800 miles below the surface of the earth lies a core of solid iron ore featuring extraordinarily high temperatures. The outer core is closer to surface, about 3,300 miles below ground, and is cooler. A mantle covers the core at approximately 1,800 miles below the surface. It is covered by the earth's crust, a much thinner layer that extends about 3 to 15 miles below the ground surface (Cassedy & Grossman, 2017). Geothermal energy emissions are the product of this extreme heat and pressure. The United States has the world's largest geothermal capacity, but it constitutes less than half of 1 percent of total energy generated (EIA, 2021).

The earth's crust varies in denseness, with the oceanic crust very thin and the continental crust very thick. Also, there are hot spots, many associated with the tectonic plates' boundaries (a source of volcanic as well as geothermal activity). Four temperature categories are associated with geothermal energy sources: high (more than 100 degrees C.), medium (90 to 160 degrees C.), low (30 to 90 degrees C.), and very low (less than

30 degrees C.). Projects to generate electricity work well only in high- and medium-grade conditions; heat for hot water and buildings can be produced in all four (Reisser & Reisser, 2019).

Geothermal energy can be exploited in three ways. First and easiest is directly using the heat, for example, by putting pipes into near-surface hot springs or hot water reservoirs, and transmitting the water to buildings or pools. The second method is to run pipes from the geothermal energy source to a surface reservoir, which is always hot. Pipes can then be run to conventional thermal turbines, to steam plants, or to power plants to generate electricity. The third and most recent way is a ground source heat pump. Ground and air heat pumps amplify energy if they use a relatively small amount of energy to move the heat, and this creates an overall efficiency, confirmed by several EPA studies (Reisser & Reisser, 2019).

Tidal and wave power. The source of tidal power is the gravitational pull of the moon twice daily on the level of seawater at the coasts. The source of wave power, indirectly, is the sun, for solar activity creates winds. Neither tidal nor wave energy currently contributes greatly to the energy profile of the United States (EIA, 2020).

Several methods have been used to exploit tidal power: a tidal barrage (a dam that the tides flow over at high tide and back out through penstocks/turbines as waters return); a tidal fence (tidal flow passes through narrow slits that turn blades in the area mixing high and low tidal water); and tidal turbines (similar to wind turbines, with the tides turning underwater propellers). These methods are limited, however, by the fact that ocean tides are cyclical and thus intermittent.

To exploit wave power, a method similar to the tidal barrage has been used. Water is channeled from large waves to a reservoir on the coast and then flows back through turbines, a system that works for large waves. Experimental work has been done to use wave turbines (like tidal turbines) in areas that have steady currents and waves; the process overall resembles that of wind farms (Reisser & Reisser, 2019).

Economic feasibility issues. Although geothermal energy has a relatively long history, most attempts to develop machinery and transmission systems for it have occurred in the last two decades. Development will be more expensive than for wind and solar renewables, and will require major support from public or private-sector organizations before projects are ready for commercialization. One example of generating electricity from geothermal is a new technology called enhanced geothermal system (EGS). It involves making a man-made reservoir that increases permeability of rocks, allowing fluids to percolate to the surface where electricity can be generated. This pilot project has a far higher cost than any technology of the mature renewables such as wind and solar (OEERE, 2012). Also, the infrastructure for geothermal will require access to transmission lines. Supporters, however, emphasize that the generating technology of this energy source could supply the electric grid with reliable energy when adequate supplies of wind and solar were not available

Environmental effects of geothermal development. While there are few harmful environmental impacts from direct use of geothermal energy for heating and warming water, power plants for electricity generation are a different matter. The plants do release a small amount of carbon dioxide into the air, and in order to remove sulfur dioxide in volcanic areas, scrubber equipment is needed. Overall, geothermal power plants produce far fewer greenhouse gases than natural gas and coal plants. The most troubling issues associated with geothermal development are contamination of water and harm to rare and precious species and their habitats.

For example, during Trump's presidency, the administration's Bureau of Land Management proposed opening more than 22,000 acres of mostly undeveloped California desert for commercial-scale geothermal power development. Called the Haiwee Geothermal Leasing Area, the acreage is situated inside a vast Desert Renewable Energy Conservation Plan (DRECP) area of nearly 11 million acres of federal land (Sahagun, 2020). The Center for Biological Diversity (CBD) condemned the proposal, asserting that it would encroach on the habitat of a state-threatened Mohave ground squirrel and a federally listed (as threatened) Mojave Desert tortoise. The group also pointed out that the proposed oil/gas development project could potentially consume hundreds of millions of gallons of water in an arid region. "The Trump administration wants to allow industrial development in conservation areas and ignore significant impacts to water resources and habitats," charged CBD senior scientist Ileene Anderson. "It's another example of the administration's ignoring science and smart planning" (Streater, 2020).

The BLM approved the leasing plan in April 2020. It defended the decision by claiming that the area could potentially spur $1 billion in geothermal power projects' investment. Environmental protests occurred during the Biden administration, when the 9th Circuit Court of appeals allowed Ormat Nevada Inc. to begin construction on its Dixie Meadows Geothermal Utilization Project. In this case, not only were vulnerable species threatened (the Dixie Valley toad), but religious practices of the Fallon Paiute-Shoshone Tribe dependent on adjacent hot springs would be adversely affected (Farah, 2022). The Center for Biological Diversity defended both the species and tribe in this case against action proposed by an energy development firm before judges selected by a Democratic president (Obama)—suggesting that not all cases in conflict follow simple political/ideological lines.

Technological challenges. Development of geothermal energy will be a test of the knowledge and capabilities of engineers, research scientists (especially geologists), and systems scientists. Most existing geothermal energy sources are found adjacent to accessible deposits of high-temperature groundwater. They are not necessarily located in areas needing energy resources. For example, in Alaska there is only one location, in the Aleutian Islands, where groundwater is hot enough to develop a major geothermal electricity power plant; but the Aleutian hot spots are sparsely populated, and power needs are greater in Anchorage and its bedroom suburbs nearly 1,000 miles away. The technology to bring geothermal power to the surface and then transport it to energy-poor regions is just now in the process of being developed (Usher, 2019).

Development costs for new technology are several times higher than for mature technology used to produce power from natural gas, oil, coal, wind, and solar and transport it to market. The completion lead times for geothermal energy are four to eight years longer than for other renewables. Too, there are significant risks to the exploration for areas rich in geothermal resources and risks as well for the production of these resources (see Muller, 2012). Private investment firms, considering these technological challenges, may decline to pursue geothermal projects. If that proves to be the case, governmental support for the sector such as subsidies and tax credits will be that much more crucial (CRS, 2021).

Wave energy development. Among the renewables, wave power is one of the most recent to be developed. The potential for harnessing ocean waves as an energy resource is vast. The DOE estimates that the technically recoverable amount is 1,170 terawatt hours, just 5 percent of which would be sufficient to provide electricity for 5 million homes. As can be imagined, however, the technological obstacles to harnessing this natural energy source are large. The first ocean wave energy project was leased by the Interior Department's Bureau of Ocean Energy Management (BOEM) in 2021. Called the PacWave South project, it lies about 6 nautical miles off the coast of Newport, Oregon, and is a cooperative venture of Oregon State University, the state of Oregon, and DOE. The Biden administration's new director of BOEM, Amanda Lefton, remarked: "Ocean waves contain a tremendous amount of energy, and this opportunity offers exciting potential to demonstrate the stability of wave energy technology and expand the nation's renewable energy portfolio" (Richards, 2021).

FURTHER READING

Cassedy, Edward S. and Peter Z. Grossman, *Introduction to Energy: Resources, Technology, and Society*, 3rd ed. New York: Cambridge University Press, 2017.

Farah, Niina. "Court clears path for Nev. Geothermal plant," *Energywire*, February 7, 2022.

Muller, Richard A. *Energy for Future Presidents: The Science behind the Headlines*. New York: W. W. Norton, 2012.

Reisser, Wesley and Colin Reisser. *Energy Resources: From Science to Society*. New York: Oxford University Press, 2019.

Rhodes, Richard. *Energy: A Human History*. New York: Simon & Schuster, 2018.

Richards, Heather. "Interior issues first-ever West Coast lease for wave energy," *E&E News*, February 17, 2021.

Sahagun, Louis. "Trump team rolls back desert protections in bid to boost geothermal energy," *Los Angeles Times*, January 23, 2020

Streater, Scott. "BLM to open 22K acres of Calif. desert to geothermal," *E&E News*, January 23, 2020.

U.S. Congressional Research Service (CRS). "The energy credit or energy investment tax credit (ITC)," *In Focus*, April 23, 2021. https://crsreports.congress.gov/product/pdf/IF/IF10479

U.S. Department of Energy, Energy Information Administration (EIA). "Tidal power," September 24, 2020.

U.S. Department of Energy, Energy Information Administration (EIA). "Use of geothermal energy," March 22, 2021.

U.S. Department of Energy, Office of Energy Efficiency & Renewable Energy (OEERE). "What is an enhanced geothermal system (EGS)?" September 2012.

Usher, Bruce. *Renewable Energy: A Primer for the Twenty-First Century*. New York: Columbia University Press, 2019.

4

Energy in Action

Electricity cannot be drilled from the ground like oil, natural gas, or other primary sources of energy. Instead, it is a force—a secondary source, derived from combustion of petroleum, coal, wood, or other primary sources. While most of the primary sources can be used just for heat and to generate motion, electricity is versatile. It can be used to turn on lights, move trucks and complex machines (your computer), and operate sophisticated electronic systems. Like the primary sources of energy, it has its special terms (e.g., kilowatt), and its supply and demand are measured on a regular basis (NAS, 2020).

The "grid"—the shorthand term often used to refer to a local, state, regional, or national electric power system—is a network of transmission lines (usually of high voltage). They are interconnected and carry electricity from the generating power sources to consumers some distance away. Substations and transformers populate the networks as well. In terms of raw numbers, there are nearly 160,000 miles of high-voltage power lines, millions of miles of low-voltage power lines, and numerous distribution transformers that connect thousands of power plants to about 145 million electricity customers. Once separated into 4,000 distinct utilities, in 2022 the grid is interconnected; it is composed of two major grids: the Eastern (including the area east of the Rocky Mountains and part of the Texas panhandle) and the Western (from the Rockies to the Pacific coast); and three minor grids: the Electric Reliability Council of Texas (ERCOT) (including most of Texas); the Alaska grid; and the Canadian power grid (which is actually integrated into the U.S. grid all along the border between the two nations) (EIA, 2019a; EPA, 2021).

Nine regional nongovernmental organizations (NGOs), which are self-regulating, are most responsible for governance of the system. Eight of the nine fall under the North American Electric Reliability Corporation (NERC), and the last is the West Coast Council, formally known as the Western Electricity Coordinating Council (WECC).

Sources of electricity in the United States. The sources of electricity in the United States are diverse, as are the technologies used to generate them. In 2020, natural gas comprised the largest source (about 40 percent) of total electricity generation. Once the largest source of electricity generation in the United States, coal was third largest in 2020, accounting for about 19 percent. Less than 1 percent of electricity was generated from petroleum, most being residual fuel oil, petroleum coke, or diesel.

Nuclear energy provided some 20 percent of the U.S. electricity profile in 2020. Renewables accounted for 20 percent as well, with the following breakdown: wind (8.4 percent of the country's electricity), hydro (7.3 percent), solar (2.3 percent), biomass (1.4 percent), and geothermal (0.4 percent) (EIA, 2021b).

The evolution of America's energy grid. As cultural anthropologist and energy scholar Gretchen Bakke has observed, "America's electric grid grew haphazardly, not radiating out from one imaginary center point but spreading instead like a pox, appearing only in spots with dense enough populations to ensure a profit. This is why for its first half century, electricity was largely an urban phenomenon" (Bakke, 2016).

The first energy grid in America was built in San Francisco in 1879, and by 1896 the Niagara Falls power plant sent power to Buffalo, New York. Large dams such as those developed along the Tennessee River during the Great Depression by the Tennessee Valley Authority (TVA) inaugurated federal government leadership of grid development. The Bonneville Power Authority was built in the Pacific Northwest in 1937, and the largest U.S. dam, Nevada's Grand Coulee Dam, began generating electric power in 1941.

During the late nineteenth and early twentieth centuries, private firms built coal- and oil-burning power plants that supplied consumers with electricity. This was a model followed through the heady days of capitalist growth; regulation of private plants was done by state governments until the federal government gained regulatory authority through the Interstate Commerce Act of 1887. But as with other areas of national life, development began in local areas and states. State governments treated providers of electricity (called utilities) as "natural monopolies," because the large costs to establish the plants and transmission wires to homes, commercial establishments, and factories reduced competition. State governments then created independent regulatory commissions, which determined entry and exit rules as well as approved rates.

Because electricity sales (especially wholesale sales) cross state lines, they are regulated by the federal government, through the Federal Energy Regulatory Commission (FERC). To complicate matters, since 1999 FERC has encouraged formation of regional transmission organizations (RTOs) and independent system operators (ISOs). This is because regional entities have much greater market power than individual units, which can save consumers billions of dollars a year. In the twenty-first century, Washington, DC, became the site of most regulatory, legislative, executive, and judicial authority over expansion of electricity and its management.

Chapter outline. Six questions organize discussion in this chapter. Questions 19 through 22 pertain to the nature of the electricity grid and discuss issues of resilience, reliability, affordability, flexibility, and sustainability. Questions 23 and 24 address both general grid security and cybersecurity concerns.

FURTHER READING

Bakke, Gretchen. *The Grid: The Fraying Wires between Americans and Our Energy Future.* New York: Bloomsbury, 2017.

Environmental Protection Administration (EPA). "U.S. electricity grid & markets." Washington, DC: June 21, 2021.

National Academies of Sciences, Engineering, and Medicine (NASEM). "Our energy sources," 2020. needtoknow.nas.edu/energy/energy-sources /electricity

U.S. Department of Energy, Energy Information Administration (EIA), "New electric generating capacity in 2019 will come from renewables and natural gas," January 10, 2019a. https://www.eia.gov/todayinenergy /detail.php?id=37952

U.S. Department of Energy, Energy Information Administration (EIA). "Electric power monthly (with data for October 2019)," November 15, 2019b. https://peakoilbarrel.com/eias-electric-power...

U.S. Department of Energy, Energy Information Administration (EIA). "Electricity explained: Electricity generation, capacity, and sales in the United States," March 18, 2021.

Q19. DOES THE CURRENT POWER GRID HAVE SUFFICIENT RELIABILITY AND RESILIENCE TO MEET EXPECTED BLACKOUTS?

Answer: No. Future large-scale blackouts, although rare, cannot be ruled out, especially if the United States fails to make significant investments in its aging grid infrastructure.

The Facts: Despite its aging infrastructure and a relatively high number of outages, the U.S. power grid meets routine power supply expectations. Nonetheless, energy infrastructure experts say that reliability and resilience will suffer and blackouts will increase in the years to come without significant improvements—especially given the increasing stresses on the grid related to climate change.

Status of the U.S. electricity grid. Most of the transformers and transmission lines in the grid are a quarter-century old; the average age of an American power plant is 34 years. The country has twice as many power plants as needed because of the inefficiencies built into the system, but so many of these plants are old and prone to maintenance issues that electricity outages are actually steadily rising on a year-by-year basis. The largest blackout in U.S. history was in 1965, when a transmission line problem cut power to 30 million people in eight Northeastern states and the Canadian provinces of Ontario and Quebec; the next largest occurred in 2003, when a two-day outage affected 50 million. The trend line went from 15 outages in 2001, to 78 in 2007, to 307 in 2011. The United States has the greatest number of outage minutes of any economically developed country, about six hours/year (excluding disasters or "acts of God," of which there were 679 from 2003 to 2012). The length of the average outage is 120 minutes, much longer than in other industrialized countries (Bakke, 2017).

Major blackouts in California and Texas. California's power problems began in 2000 after deregulation of the state's electricity system. The state grid manager initiated rolling blackouts to balance power needs of different regions, and also the state urged consumers to limit electricity use during peak times. These measures allowed continued use of elderly infrastructure, which was a factor said to have caused forest fires in 2017 and 2018, and liability issues leading to the bankruptcy of the state's largest utility, Pacific Gas & Electric Co. (PG&E). The tightening of electricity supplies left the state at least 500 megawatts short of electricity at the time of the August 2020 heat wave. The resulting rolling power outages were subsequently cited as a factor in fueling the fires in Southern California that ultimately burned about 4.2 million acres (Mulkern, 2021).

Different extreme weather conditions—unprecedented cold temperatures and ice—caused major electricity outages in Texas and other states in late February 2021. Cold weather had been forecast, but still more than 4 million Texas homes and businesses were left without power (and 12 million lacked access to clean drinking water); at least 200 people died during the storm, and household utility bills soared; power costs increased manyfold. Said University of Texas Austin research associate Joshua Reynolds: "[The frigid weather is] such a black swan event. . . . It is taxing every single piece of the system at the same time" (Klump et al., 2021a).

The International Energy Agency (IEA) opined that the Texas outages were 500 times higher than those in California (IEA, 2021). Lights (and heat) were on for most Texans within a few days, but the recovery costs were astronomical. Many Texans were on fixed-price utility plans, which protected them from market swings. Others had market-based plans, and when the wholesale price of electricity shot up, these customers received bills greater than $5,000 (and some as much as $10,000). The human costs included an 11-year-old boy whose death was attributed to hypothermia when the family home lost power (Klump et al., 2021b).

Management of the grid and its expansion. The North American Electric Reliability Corporation (NERC) is in charge of managing reliability standards for grid operation under a plan approved by the Federal Energy Regulatory Commission (FERC). Enforcement of reliability standards is largely in the hands of the NERC. It ascertains whether adequate electricity is available in the major and minor grids; if not, grids with insufficient power can transfer from surplus reserve grids. In the Texas case, the Electric Reliability Council of Texas (ERCOT) had isolated the state's energy grid from Eastern and Western regional grids for decades, perhaps in the belief that its vast coal and natural gas resources would suffice to meet state needs. The February 2021 storm proved otherwise, and immediate transfers of power from other networks did not occur. Several other factors contributed to the Texas blackout as well, including low reserves (coal, natural gas) of backup fuel; power plants that had never been winterized; and critical infrastructure and machine parts that were not protected from freezing [Lee & Klump, 2021]). The Texas case is just the most recent example of ambiguity regarding governance of the grid. As a 2021 National Academies of Sciences, Engineering, and Medicine report on U.S. electric power pointed out, "much as no single entity plans the internet, no single entity is responsible for planning the future of the power system" (NASEM, 2021). The 50 states are not uniform in their regulation of utilities, however, which creates even more confusion and uncertainty about management and expansion of the nation's energy grid.

The Texas blackout postmortem showed how difficult it can be to arrive at solutions for shoring up the nation's energy grid, as different constituencies blame different factors for its current vulnerable state. Republican governor Greg Abbott of Texas, a close ally of the state's huge oil and gas industry, blamed frozen wind turbines for the outage, even though the state's natural gas infrastructure also was frozen and supplies a larger portion of Texas power. The economic advisor of former President Trump blamed the outage on President Biden's climate change policies, even though Biden had been in office for less than two weeks (Zhao, 2021). A review of PUC documents, meanwhile, revealed serious regulatory failure. For example, the influential natural gas lobby successfully blocked attempts by regulators to winterize the state's grid (Klump et al., 2021c). At a public hearing on the catastrophe, PUC chair Peter Lake said, "We are taking a blank-slate approach for a full overhaul and redesign of this market to drive reliability." Governor Abbot, however, made it clear that he did not want any changes to policy concerning gas, coal, and nuclear power, that would make their treatment by the state inferior to that of renewables (Klump & Lee, 2021).

One year after the blackouts, large gaps remained in the regulatory system. For about 40 percent of the pipeline and storage sites the state had deemed critical, records of the Texas Railroad Commission (TRC) revealed that company officials either had not conducted a winterization test or did not know if one had been done. Critics pointed out that the TRC had not yet written winterization standards for gas wells, pipelines, and storage facilities. Luke Metzger, executive director of Environment Texas, said lack of clean standards made it hard to evaluate the effectiveness of the inspection campaign: "I worry that might be more of a public relations stunt than a credible regulatory effect" (Lee & Soraghan, 2022).

Tests of grid resilience; black start. Congress also has devoted attention to blackouts and grid resilience. For example, in 2018 the U.S. Senate Committee on Energy and Natural Resources held a hearing on "black start" generation capacity. (A broader definition of black start is restoring part of a grid to operation without relying on an external network; instead, it is restarted by an auxiliary power source, such as a small diesel or gas-fired generator or battery.) Committee chair Lisa Murkowski (R-AK) presented a scenario of a blackout spanning from Maine to Florida, and all the way to Minnesota and Louisiana, leaving hundreds of millions without power. Some experts thought that American utilities could bounce back from a major outage. An energy laboratory director, on the other hand, said that re-energizing the grid after a massive outage was an "intricate and multifaceted endeavor fraught with potential unforeseen technical challenges." Instead of a jump-start to the system, a more gradual approach would be needed (Sobczak, 2018).

In spring 2020, NERC evaluated the grid's risk profile because of potential workforce disruptions, supply chain interruptions, and increased cybersecurity threats. The report's conclusion was rosy and did not identify "any specific threat or degradation to the reliable operation of the bulk power system" (NERC, 2020). These different perspectives revealed what authors of an ICF International report called a "resilience gap": the energy industry, electric utilities, government regulators, the public—all had different agendas, making consensus extraordinarily difficult to reach. Such a consensus would be essential to protect critical infrastructure against the effects of climate change (Behr, 2021).

Biden administration views and priorities Both as a candidate and as president, Biden placed a heavy emphasis on dealing with the already visible effects of climate change—rising temperatures, extreme storms, rampant wildfires, and rising sea levels. On the campaign trail, he repeatedly mentioned the need to safeguard critical energy systems against damage and harden the backup infrastructure against the malign impacts of a warming planet. He also asserted that the country needed to ramp up investment in green energy sources. The contrast was clear between Biden's policy priorities and the Trump administration's America First ideology, which wholeheartedly embraced fossil fuels and dismissed concerns about climate change.

Biden acknowledged the initial higher costs of newer forms of energy (wind and solar) and of electric vehicles, which would require subsidies from the federal and state governments as well as major investments of corporations. The second of three waves of new spending, the American Jobs Act, was part of stimulus legislation early in the Biden presidency. The third wave was unprecedented, a $2.2 trillion infrastructure plan. In a major speech to Congress in April 2021, the president said the nation's transportation, energy, and electricity sectors were broken and required major overhauls—to continue the energy transition, to address climate change, to restore competitiveness with other nations, and to create jobs. He termed it the "blue collar blueprint to build America" (Clark, 2021).

The final effort of the Biden administration was mired in intra-party controversy in 2022, the year of mid-term elections for the Congress. The large caucus of progressive Democrats sought to revive the "Build Back Better" plan to direct hundreds of billions for clean energy technology to combat climate change. Moderates, such as Joe Manchin (D-W.Va) and Kyrsten Sinema (D-AZ) insisted that the package be fully paid for across the 10-year budget window. Observers were not optimistic that a compromise could be reached (Dillon & Sobczyk, 2022).

FURTHER READING

Bakke, Gretchen. *The Grid: The Fraying Wires between Americans and Our Energy Future.* New York: Bloomsbury, 2017.

Behr, Peter. "Report: 22 power lines could boost renewables by 50%," *E&E News*, April 27, 2021.

Clark, Lesley. "Biden lays out 'blue-collar blue print' to transform energy," *E&E News*, April 29, 2021.

Dillon, Jeremy and Nick Sobczyk. "4 issues to watch as Dems eye reconciliation return," *Energywire*, April 25, 2022.

International Energy Agency (IEA). "Severe power cuts in Texas highlight energy security risks related to extreme weather events," February 18, 2021.

Klump, Edward, Peter Behr, and Mike Lee. "Bitter cold overwhelms grid, leaves millions in dark," *E&E News*, February 16, 2021a.

Klump, Edward, Lesley Clark, and Mike Lee. "'Heads will roll': Grid crisis sparks political firestorm," *E&E News*, February 22, 2021b.

Klump, Edward, Mike Lee, and Carlos Anchondo. "Documents reveal natural gas chaos in Texas blackouts," *E&E News*, May 20, 2021c.

Klump, Edward and Mike Lee. "Texas plans 'monumental' electricity overhaul," *E&E News*, July 14, 2021.

Lee, Mike and Edward Klump. "'Train wreck,' energy CEOs call for Texas grid overhaul," *E&E News*, February 26, 2021.

Lee, Mike and Mike Soraghan. "Documents show major gaps in Texas gas inspections," *Energywire*, February 7, 2022.

Mulkern, Anne. "Blackouts trigger Calif. energy storage boom," *E&E News*, July 13, 2020.

Mulkern, Anne. "What Calif. blackouts reveal about U.S. grid," *E&E News*, March 16, 2021.

National Academies of Sciences, Engineering and Medicine (NASEM). "The future of electric power in the United States." Washington, DC: January 2021. https://nap.nationalacademies.org/catalog/25968/the-future-of-electric-power-in-the-united-states

North American Electric Reliability Corporation (NERC). "Special report: Pandemic preparedness and operational assessment: Spring 2020." https://www.nerc.com/pa/rrm/bpsa/Alerts%20DL/NERC_Pandemic_Preparedness_and_Op_Assessment_Spring_2020.pdf

Sobczak, Blake. "Grid planners put 'black start' technology to the test," *E&E News*, November 13, 2018.

Zhao, Christine. "Larry Kudlow suggests Texas power outages 'consequence' of Biden's presidency," *Newsweek*, February 2021.

Q20. WILL THE ELECTRICITY GRID OF THE UNITED STATES HAVE SUFFICIENT CAPABILITY BY 2030 TO MEET EXPECTED NEEDS FOR POWER IN THE NEXT GENERATION?

Answer: Unknown, but the longer the United States goes without modernizing its energy grid, the less likely it is to be able to meet energy demands in the future

The Facts: Serious planning for energy grid protection began only in 2009 with the release of a Department of Energy (DOE) program called the Grid Modernization Initiative (GMI). The GMI includes a portfolio of activities crafted to help develop a cost-effective roadmap to a reliable, resilient, secure, and sustainable grid at a cost affordable to the nation and individual consumers. It is like an architectural design with new concepts, tools and technologies. The DOE is the keystone agency, supported by the national laboratories under the name Grid Modernization Lab Consortium.

The GMI brought together a consortium of 14 DOE national laboratories and regional networks to help develop and put the multiyear program plan into effect. An important aim was to enhance coordination between DOE offices focused on different types of energy, as well as DOE coordination with other federal agencies, legislators, regulators, utilities, consumer groups, and the like (DOE, 2021). All this is to be done when the demand for renewables and natural gas to generate electricity likely will increase (Klump, 2021). The following examples demonstrate the difficulty of achieving coherence.

Conflict among priorities. It was not clear that grid modernization was a top priority of the Trump presidency. Energy Secretary Perry's attempts to bolster struggling coal and nuclear plants (at that time the second- and third-largest sources for generating electricity) won little support in his department, the White House, or in the National Security Council. The Federal Energy Regulatory Commission (FERC) could make emergency purchases, but when Perry asked the agency to buy electricity from power plants, it declined (Northey, 2018). Finally, the idea of grid modernization dated from the Obama administration, many policies of which were repudiated by Trump simply because of their origins.

The Biden presidency's emphasis on climate change adaptation and renewables contrasted sharply with that of his predecessor. Studies indicated that significant decarbonization of the economy would require a large expansion of electric transmission, but transmission siting and approval had been controlled by the states. The 2005 Energy Policy Act allowed federal intervention if the grid is heavily congested, which it had been. When, shortly after his inauguration, Biden directed federal agencies to speed up clean energy and transmission deployment as part of its grid modernization efforts, it received backlash from red states (Tomich, 2021).

Technological change; planning and regulatory responses. Modernizing the grid by adding new energy sources to meet demand (with a focus on renewables, especially solar and wind) required technological solutions to three questions of access as well as planning and regulatory reform. Solar and wind receive the sun's bounty intermittently, and batteries make energy flow stable. Technological innovations have reduced the size of batteries while increasing their storage capacity, but these renewables are still not on a par with hydro/ nuclear power sources, which have significant on-site storage capabilities.

Another problem has been development of a means to bring electric power to vulnerable communities that are far from power sources or particularly exposed to interruptions in service. In recent years, engineers have developed mini-grids to bring power to people when natural disasters (fires, floods, hurricanes) close down the grid. The formal name, micro-grid, is used to describe a collection of energy generators crafted to serve nearby customers, such as a large apartment complex or university campus. The power for micro-grids often is supplied by generators, but solar arrays with battery storage have become popular. A number of entities have become players in micro-grid development, including Tesla, utilities such as Pacific Gas & Electric, and global firms such as Siemens AG and Caterpillar (Ferris, 2020).

Other difficulties in the modernization process pertain to design, planning, and regulatory practice. First, in addition to the congestion problems mentioned, energy sources distant and remote—for example, deserts, mountain tops, offshore waters—are not securely linked to the grid by high-voltage transmission lines (Richards, 2021). Second, planning is proceeding very slowly on the macro-grid (the interregional high-voltage direct current [HVDC] project), a planned infrastructure upgrade to connect wind power generated in the Midwest to high-demand areas on the East coast. It will bury transmission lines underground to avoid conflict with landowners (Tomich, 2021). Third, renewable power advocates, such as the American Council on Renewable Energy, blame the regulatory system itself. They note that building high-voltage power lines takes 7 to 10 years, which is too long a time frame to meet environmental (especially climate) challenges in the near-term (Willson, 2021).

In late 2020, the FERC adopted Order 2222, which permitted "distributed energy resources" (DERs) such as rooftop solar, as well as a range of other sources (e.g., storage batteries) to trade in wholesale energy markets (Skibell, 2020). In mid-2021, new FERC chair Richard Glick asked for consideration of further reforms in order to ease the backlog of applications by renewable energy companies eager to connect to major grids across the country. The primary issue to regulators has been determining a fair method to pay for upgrading the grid (Tomich, 2021). (In addition, partisan conflict on the FERC influenced the commission's oversight of natural gas projects regarding pipelines' GHG emissions [Willson, 2022].)

Under the "beneficiary pays" rule adopted by FERC in 2011, renewables bear almost the full cost of increasingly expensive upgrades (such as adding new or making upgrades to the transmission lines) in the Midwest and Great Plains, which they find unreasonable. They contend there are broad benefits to the grid as a whole for adding solar and wind. As a result, low-cost renewables don't pay for new lanes in the system, but are stuck in long interconnection queues. (The ratepayers [consumers] of solar and wind especially object to this.)

Within the first 100 days of his administration, Biden moved to restructure both the electricity sector and the auto industry. He proposed a decarbonized electricity grid by 2035 and a carbon-free economy by 2050. These were inspiring (albeit largely aspirational) goals. A Princeton University report concluded that reaching the 2035 target would require 50 million electric vehicle purchases and installation of 3 million public charging posts. Wind and solar energy generation would need to increase fourfold. Meeting all these goals will require the United States to, in the words of former Obama administration energy secretary Munoz, "innovate like hell" (Behr & Clark, 2021).

Financial feasibility. Cost is an obvious impediment to implementation of the GMI multiyear program plan. In 2015, the Edison Electric Institute estimated that the industry would need to spend between $1.5 and $2 trillion from 2010 to 2030 just to maintain the reliability of the service. The replacement value for power plants alone, which account for a little more than half of the cost of the grid, would be $2.7 trillion. In 2017, these replacement costs were closer to $5 trillion (Paraskova, 2017).

The electricity infrastructure is capital intensive, and costs of replacing it are not easy to ascertain because of fragmented ownership. Most of the power plants, transmission lines, substations, and transformers in the system have been financed by private investors, who are willing to invest because of confidence that state PUCs and the FERC will set electricity usage rates that will enable them to profit on their investments. Financing from the federal and state governments has been a smaller part of the mix (Willrich, 2017). As part of the effort to spur development of renewables, Biden's Department of Energy promised $8.15 billion in loans from its office as well as from the Western Area Power Administration to expand high-voltage transmission lines, continuing the modernization of the grid. The Biden administration also asked state governments to treat clean energy projects as "utility" investments (Behr, 2021).

The bipartisan infrastructure bill passed by the Senate in early August 2021 was the largest in U.S. history. It included $550 billion in new infrastructure spending (of an overall multiyear total of $2.2 trillion). The bill was a mixed bag. It included major investments in clean energy research; annual funding (starting at $45 million) for energy efficiency projects; and $7.5 billion for EV chargers (only half of what the Biden administration sought). Transportation funding was increased by $7.5 billion for clean buses and ferries, but the appropriation language also allows some funding for fossil fuel-dependent investments. Finally, the infrastructure package did not contain renewals of existing investment tax credits and production tax credits, although these could be included in later legislation (Cahlink & Dillon, 2021). A critical analysis of the infrastructure bill by the International Renewable Energy Agency (IRENA) suggested that its investments in new fossil fuel projects would "lock-in uneconomic practices, perpetuate existing risks and increase the threats of climate change" (Bond, 2022).

FURTHER READING

Bakke, Gretchen. *The Grid: The Fraying Wires between Americans and Our Energy Future*. New York: Bloomsbury, 2017.

Behr, Peter. "DOE unveils grid plans to unlock renewables," *E&E News*, April 28, 2021a.

Behr, Peter. "Wanted: 'Superhuman' AI to master a greener grid," *E&E News*, August 23, 2021b.

Behr, Peter and Lesley Clark. "Biden lays foundation for energy overhaul," *E&E News*, April 26, 2021.

Bond, Camille. "How expensive is CO2-free energy? What 2 reports say," *Energywire*, March 30, 2022.

Cahlink, George and Jeremy Dillon. "Energy winners and losers in the bipartisan infrastructure package," *E&E News*, August 10, 2021.

Ferris, David. "Tesla microgrids spread as U.S. grid 'gets worse and worse,'" *E&E News*, October 28, 2020.

Klump, Edward. "Renewables and natural gas: What's ahead in 2021," *E&E News*, January 12, 2021.

Klump, Edward, Kristi Swartz, and Jeffrey Tomich. "Coronavirus and electricity: 4 takeaways from CEOs," *E&E News*, May 20, 2020.

Northey, Hannah. "Poorly articulated grid plan stalls," *E&E News*, October 16, 2018.

Paraskova, Tsvetana. "US electric grid could cost $5 trillion to replace," *Oil Price*, May 23, 2017.

Richards, Heather. "FERC pressed on connecting offshore wind to the grid," *E&E News*, May 17, 2021.

Skibell, Arianna. "'Game-changer': FERC order opens door for renewables," *E&E News*, September 18, 2020.

Tomich, Jeffrey. "Can Biden transmission order avoid state backlash?" *E&E News*, February 11, 2021a.

Tomich, Jeffrey. "Midwest transmission morass: A 100% clean power warning?" *E&E News*, March 25, 2021b.

Tomich, Jeffrey. "Glick eyes transmission reform amid renewables bottle-neck," *E&E News*, May 19, 2021c.

U.S. Department of Energy. *Grid Modernization Multi-Year Program Plan*. Washington, DC, 2021. https://www.energy.gov/downloads/grid-modernization-multi-year-program-plan-mypp

Willrich, Mason. *Modernizing America's Electricity Infrastructure*. Cambridge, MA: MIT Press, 2017.

Willson, Miranda. "Poor transmission planning hamstrings renewables—report," *E&E News*, April 6, 2021.

Willson, Miranda. "FERC hearing: Gas fights, Manchin and a 'snowball effect'," *Energywire*, March 3, 2022.

Q21. IS THE U.S. POWER GRID SUSTAINABLE?

Answer: Perhaps, for the near term; but it is not likely to be sustainable into the mid- and the long-term if demand for electricity increases substantially. The DOE's Energy Information Administration predicts an increase in demand of 1 percent annually until 2050. The existing U.S. power grid can meet this modest increase if there are continued increases in energy efficiency (such as phasing out coal and replacing it with natural gas and renewables). However, climate change adaptation (e.g., heat waves and other extreme weather events) would increase overall costs.

The Facts: The concept "sustainable" means that at least as much of a particular resource (e.g., food, water, housing, transportation, or electricity) will be available for future generations as is available for us now. In this question, the focus is on whether America's existing power grid provides *access* to a sufficient amount of energy to meet consumer needs, both now and in the future.

The power grid does supply sufficient electricity to meet the needs of all consumers who can afford it (discussed in Q22), but the current system will not be adequate for future generations. In light of this reality, the National Academies of Sciences, Engineering, and Medicine has called for the development of a high-voltage interstate power line network as the foundation for a low-carbon national power grid (Behr, 2021; National Academies, 2021). Making such schemes a reality, however, involves daunting challenges. In addition to technological and regulatory challenges raised in Q20, other obstacles frustrate efforts to make the nation's energy grid more sustainable and reliable, including state/local opposition to a national system and a highly polarized political environment that makes it enormously difficult to pass vital legislation.

Need for a high-voltage electricity transmission system. The U.S. power grid has multiple parts, and only with great effort has it been possible for them to function sufficiently well in the absence of a national high-voltage transmission system. Energy conservation has made the overall system more efficient, reducing temporarily the need to increase facilities for transmission. For example, increased efficiencies have reduced average household energy expenditures from 5.1 percent of all energy usage in 2009 to 4 percent in 2019. Yet increased abundance of lower-cost fuels has also "required larger and smarter delivery networks" (Zerrenner, 2020).

WIRES, an advocate for power line expansion, coauthored a report on sustainability with Scottmadden (an energy management consulting firm). It analyzed region by region the opportunities for connecting surplus renewable energy areas, such as the Midwest, to large metropolitan areas likely to experience energy deficits in the future. A related study by the National Renewable Energy Laboratory (NREL) had a similar objective. It argued that the three, now separated grids (Eastern, Western, Texas) should be connected into a single, integrated network capable of generating much cheaper and cleaner electricity (Behr, 2018; Uhlenhuth, 2018).

Joining these advocates for expansion of power transmission lines was an unlikely ally, the U.S. Climate Alliance. This is a bipartisan coalition of nearly half the American states that seeks to reduce greenhouse gas emission, develop clean energy finance, and promote power-sector modernization. Its campaign to construct new high-voltage lines would accelerate the energy transition from fossil fuels to renewables (U.S. Climate Alliance, 2021). The need to harness renewables is particularly important, as they account for three-quarters of utility-scale electricity-generating capacity in 2021 nationally (EIA, 2021; Klump, 2021a). Another argument advanced to justify construction of high-voltage lines was the risk to the electricity system from water shortages. Inadequate water supplies endanger fossil fuel power plants and nuclear reactors, both of which require large reserves of water to operate. Approximately 70 percent of American power plant capacity requires cooling waters, most coming from surface waters (Northey & Behr, 2021).

The process of developing a high-voltage transmission system could begin with a presidential administration or the Congress; alternatively, it could emanate from the private sector (which might be easier in times when Congress is divided). Reaching agreement on such a system, however, will require both Republicans and Democrats to compromise. A 2021 report from the Clean Air Task Force (a center-left nonprofit) and the Niskanen Center (a center-right advocate of property rights/free markets) reflected this need. Their report urged a comprehensive approach to planning and building national transmission lines, and quick movement to carbon-free power to avoid "devastating impacts of climate change" (Willson, 2021).

Opposition of state and local governments. The American system of federalism often places states and local governments in competitive relationships with one another on grid issues. Some regions of the country, such as parts of the West and Southeast, are nonetheless resistant to the idea of being part of a national grid (Klump, 2021b). As mentioned in Q19, Texas has its own system, mostly isolated from the rest of the United States. Transmission lines that cross state lines have become increasingly difficult to site and build as a result.

The report by Scottmadden and WIRES complained about the lack of a national perspective on America's energy needs on the part of these states and cities. It said their policies "continue to stymie transmission development" because the processes they use to identify locations for projects and to consider and implement construction permits take too much time. Many states, cities, and counties are more concerned with keeping employment high and companies successful in their own energy sectors. For example, the National Rural Electric Cooperative Association transported hundreds of members to Washington, DC, to protest President Biden's proposal to create a zero-carbon grid by 2035. Members of this association produce only about 5 percent of the electricity generated in the United States, but they are spread across 55 percent of the country's lands. They robustly defend decentralized grids, without which they would lose members (Ferris, 2021).

FURTHER READING

Allan, Jens, ed. *The State of Environmental Governance 2019.* Winnipeg, Manitoba, Can.: International Institute for Sustainable Development (IISD), Earth Negotiations Bulletin, February 2020.

Behr, Peter. "DOE searches for certainty in the grid's future," *E&E News*, November 16, 2018.

Behr, Peter. "National Academies offers stark warnings for shifting grid," *E&E News*, February 26, 2021.

Ferris, David. "Electric co-ops: Biden clean power plan 'overly ambitious,'" *E&E News*, April 22, 2021.

Jiusto, Scott and Stephen McCauley. "Assessing sustainability transition in the US electrical power system," *Sustainability*, Vol 2, no. 2 (2010): 551–75. https://doi.org/10.3390/su2020551

Kenward, Alyson and Urooj Raja. *Blackout: Extreme Weather, Climate Change and Power Outages.* Princeton, NJ: Climate Central, 2014.

Klump, Edward. "Renewables and natural gas: What's ahead in 2021," *E&E News*, January 12, 2021a.

Klump, Edward. "Western U.S. grid plan could remake renewables," *E&E News*, July 2, 2021b.

National Academies of Sciences, Engineering, and Medicine (National Academies). "The future of electric power in the United States." Washington, DC: January 2021.

Northey, Hannah and Peter Behr. "Severe heat, drought pack dual threat to power plants," *E&E News*, June 28, 2021.

Uhlenhuth, Karen. "Transmission study points to potential from overcoming grid seams," *Energy News*, August 1, 2018.

U.S. Army Corps of Engineers. *2019 Sustainability Report and Implementation Plan*. June 30, 2019.

U.S. Climate Alliance. "U.S. Climate Alliance Releases 2021 Annual Report, Details. www.usclimatealliance.org/publications/2021/12/15/2021-annual-report

U.S. Department of Energy, Energy Information Administration (EIA). "Electricity explained: Electricity generation, capacity, and sales in the United States," March 18, 2021.

U.S. Executive Office of the President. *National Electric Grid Security and Resilience Action Plan*. Washington, DC: December 2016.

Zerrenner, Kate. "Sustainable Energy in America, By the Big—No, HUGE—Numbers," *Energy & Environment*, February 12, 2020.

Q22. IS ELECTRICITY AFFORDABLE FOR MOST AMERICANS?

Answer: Yes, most (about four-fifths of) Americans can pay for the electricity they need, but a minority cannot do so without government assistance.

The Facts: The U.S. power grid is being modernized, and improvement will be costly. Staged over a generation, the national costs will be bearable. However, it has been estimated that nearly one in five Americans may not have the financial means to secure their electricity needs in the future without assistance. Costs of grid modernization reflect the decisions and priorities of leaders, and fossil fuel power sources (favored in the Trump administration) have become less important in the Biden administration, which is more focused on modernizing America's power grid with renewable energy sources. Oversight of the electric system is conducted primarily by the Federal Energy Regulatory Commission

(FERC), and its mandate is to reduce costs while improving efficiency of energy service in the United States.

Modernization of the U.S. electricity grid. Energy expert Mason Willrich accurately described America's existing electric infrastructure as "a network of networks connecting generators to loads (consumer demands) over a patchwork of 66 geographic balancing authorities. . . . (They range in size) from very small to very large and multistate within three very large interconnections including Eastern, Western, and Texas" (Willrich, 2017).

In addition to being a patchwork, the grid is old. The average age of power plants is older than 30 years, and that of dams greater than 40–50 years, which is considerably beyond their shelf lives. Improvements in generators have been incremental and piecemeal as well. In many parts of the country, for example, communities are dependent on generators that still run on fossil fuels.

The grid is also incomplete in several respects. The need for a high-voltage electricity transmission system is clear, and it is becoming more urgent with each passing year (see Q21). Planning for establishing a truly national grid has been undertaken and implementation is under way, but will be done in stages that could drag on for decades in the absence of political consensus on priorities. Another example concerns electric batteries and battery storage capability. Narratives about electric vehicles (Q7) describe uncertainty about battery storage capability for long-distance travel, yet within the last decade, energy storage deployment has expanded, and solar-plus storage projects have demonstrated their commercial viability (a solar-plus storage system is a battery system that is charged by a connected solar system (e.g., a photovoltaic one).

A generation ago, travel by electric vehicle (EV) from Anchorage to Fairbanks, Alaska—a distance of 359 miles—was thought to be impossible. Today, the route includes an EV charging station, enabling electric vehicle owners to travel from Anchorage to Fairbanks without worrying whether they have enough juice. Originally, it took an hour to charge a car to drive for 25 miles; by 2021, though, that same hour of charging produced enough juice to drive that same car 225 miles (Capps, 2021). At the outset in 2017, the charge time was many hours; in 2021, it is about 45 minutes; in 2030, some experts believe it might not take any longer to recharge an electric vehicle than it does to fill a tank with gasoline today.

Grid spending priorities shift from administration to administration. Estimates of the cost of replacing the U.S. electric grid reach as high as $5 trillion, but about two-thirds of the infrastructure is privately owned or held by cooperatives. In the period from 2010 through 2018 (the last year for which complete data are available), privately owned utilities invested $170 billion to support buildout of the transmission grid (Zerrenner, 2020). Going forward, these private utilities and cooperatives will undoubtedly lobby for federal loans and other incentives to help them modernize. Meanwhile, investors have nearly doubled their rate of spending on infrastructure in recent years. One example is Breakthrough Energy Ventures (BEV), a group of investment banks, philanthropies, and corporations organized by Bill Gates in 2015 that is focused on using technologies to attain net-zero carbon emissions by 2050 (Fialka, 2020).

As with other areas of infrastructure such as roads, bridges, and treatment plants, government spending on electric grid infrastructure has not kept up with basic maintenance needs, to say nothing of modernizing systems to meet twenty-first-century demands. The Trump administration primarily championed the interests of fossil fuels and related infrastructure (roads, railroads, pipelines), and its last executive budget proposal reduced non-defense discretionary expenditures (including for renewables) by 7 percent (Cahlink, 2020). Congress has extended clean energy tax breaks and increased R&D funding in the national laboratories and the Advanced Research Projects Agency-Energy (ARPA-E) (Factbook, 2020). Funding for grid modernization projects rebounded, however, as the incoming Biden administration gained funding for jobs in energy and large infrastructure projects (see Q20).

Costs for individual electricity consumers. Most Americans have affordable electric bills. Prices for household electrical commodities and equipment have dropped sharply. A good example is the light-emitting diode (LED) lightbulb. In 2012, the price of an "A-type" LED was $37 per thousand lumens produced, but by 2017 the cost had fallen by 80 percent (EIA, 2019). Use of LEDs increased from almost zero in 2000 to 1.1 billion units in 2018. Furnaces, freezers, air conditioners, clothes washers, refrigerators, TVs, and other home appliances—all showed similar efficiencies (Factbook, 2020). Altogether, increased energy conservation efforts reduced average household energy expenditures from 5.1 percent in 2009 to 4 percent in 2019 (BloombergNEF, 2019).

However, millions of Americans struggle to pay for the electricity they need. Most households struggling to pay their electric bills receive public assistance, such as Temporary Assistance for Needy Families (TANF), Supplemental Security Income (SSI) benefits, housing assistance, or other aid. Yet electricity costs are not adjusted by income. In 2019, the Citizens Utility Board, a utility NGO in Chicago, investigated income disparities of electricity consumers. Researchers collected electricity meter data of customers and correlated them with census data from different neighborhoods. They found that affluent households used more energy than low-income customers, especially during peak hours of usage (e.g., for air conditioning on hot days) (Tomich, 2019).

Supporting this interpretation is a recent working paper by a doctoral economics student at UC Berkeley pointing to costs for African American renters and homeowners that are several hundred dollars higher than for White residents (Klump, 2020). These findings are attributable to several factors, including: 1) The poor pay a larger part of their income for electricity (10 to 20 percent) as compared to the rich (who pay 1.5 to 3 percent). 2) The poor and especially Black residents have a greater housing density than the rich. They have larger families, and more people living in smaller-size units. (Sometimes this situation is the result of segregation.) 3) The poor tend to live in older housing stock, and there are fewer incentives for developers to build housing that is more energy efficient. (Subsidized rents also reduce developers' incentives to reduce energy costs.) 4) Finally, unlike most White families, Black families often are required to sign long-term contracts for electricity and heating services, and these reduce their ability to shop for better prices.

Among the Biden administration's other stated priorities was to provide new funding for programs to help poor people pay utility bills and to address long-standing environmental justice issues, such as the disproportionate impact of pollution and other forms of environmental degradation on poor communities of color (Brugger, 2021). In 2021, President Biden's home state of Delaware became the first to establish a committee to implement the "Justice40 Initiative." This environmental justice policy would allocate 40 percent of climate and energy program benefits to disadvantaged communities (Willson, 2022).

Oversight by the Federal Energy Regulatory Commission (FERC). This independent regulatory commission regulates public utility commissions (PUCs) in the United States, and thus has substantial influence over the operation of the power grid. Like most other regulatory commissions, it has an uneven number of commissioners (up to five, representing both political parties), who are appointed by the president for staggered terms. Congress assigned security of the interstate power grid to FERC in the 2005 Energy Policy Act.

In 2010, FERC became more directly involved in grid transmission planning by its adoption of Order No. 1000 (FERC, 2011). This order required public utility units to participate in interregional transmission plans (e.g., to exchange power with regions outside their boundaries). It asked public utilities to estimate benefits of transmission line plans and prioritize them. It also allowed for-profit utilities to build new lines or upgrade existing ones, and to recoup their investments by increasing electric rates, as public utility commissions are permitted to do so (Access Intelligence, 2012). The energy consulting firm Scottmadden analyzed five years of data under the new order. It found that the establishment of new processes had not led to increased cost-effectiveness or efficiencies (Lyons & Messick, 2016).

For the first time in its existence, the FERC joined with the National Association of Regulatory Utility Commissioners (NARUC) in 2021 to explore new ways to determine the costs of transmission upgrades and expedite the way energy projects connect to the grid. NARUC represents state utility regulators, who have had frosty relationships with FERC regarding state subsidies for renewables and their impact on pricing in key grid markets. Renewable energy advocates applauded the change because it potentially expedites connecting new wind, solar, and other clean energy resources to the grid (Willson, 2021).

FERC is relatively new to the oversight role, and it brings an inclusive perspective to utilities management. Energy experts hope that its emphasis on policy reforms and market changes will ultimately result in more affordable electricity for consumers.

FURTHER READING

Access Intelligence. "Power, FERC Rule 1000: What does it mean?" June 30, 2012.

Aton, Adam. "Exclusive: White House details environmental justice plans," *E&E News*, July 20, 2021.

Brugger, Kelsey. "'Go full force': High hopes for environmental justice panel," *E&E News*, March 30, 2021.

Cahlink, George. "Trump proposes deep energy, environmental cuts," *E&E News*, February 10, 2020.

Capps, Kris. "Northernmost EV fast charger opens near Cantwell," *Fairbanks Daily News-Miner*, August 29, 2021.

Federal Energy Regulatory Commission (FERC). "Order No. 1000— Transmission planning and cost allocation," 2011.

Fialka, John. "Roaring '20s? Not enough for energy innovation," *E&E News*, December 2, 2020.

Klump, Edward. "Black households pay more for energy than White ones," *E&E News*, June 25, 2020.

Kuckro, Rod. "CEOs see technology, customers helping to cut power demand," *E&E News*, November 13, 2018.

Lyons, Cristin and Brian Messick. *FERC Order No. 1000: Five Years On.* Scottmadden, June 2016. https://www.scottmadden.com/insight/ferc-order-no-1000-fiveyear

Sliman, Kevin. "Shedding light on the rules that govern the U.S. power grid." *PennState News Release*, February 11, 2020.

Tomich, Jeffrey. "Chicago electricity study says poor subsidizing rich," *E&E News*, June 9, 2019.

U.S. Department of Energy, Energy Information Administration (EIA). "Electric power monthly with data for October 2019," December 2019.

Willrich, Mason. *Modernizing America's Electricity Infrastructure*. Cambridge, MA: MIT Press, 2017.

Willson, Miranda. "'Big step' on transmission? High hopes for FERC task force," *E&E News*, June 23, 2021.

Willson, Miranda. "Delaware offers litmus test for Biden's EJ plan," *Energywire*, February 3, 2022.

Zerrenner, Kate. "Sustainable Energy in America, By the Big—No, HUGE—Numbers," *Energy & Environment*, February 12, 2020.

Q23. CAN THE U.S. ENERGY GRID BE PROTECTED AGAINST ENVIRONMENTAL AND SECURITY THREATS AND PRESSURES?

Answer: The pressures and risks can be reduced but not eliminated.

The Facts: Because the electricity grid is composed of hundreds of components, thousands of wires, and millions of contact points in cell phones, among other elements, its exposure to environmental hazards is very high. Three topics help clarify the nature of the risks and responses to them: physical infrastructure including telecommunication devices; catastrophic atmospheric events; and extreme weather events accelerated by climate change.

Risks to physical infrastructure of the power grid. During the 2019–20 period of robust commerce and trade negotiations between China and the United States, elements of the power grid attracted scrutiny, including transformers, drones, and cell phones. In each area, China produced products that have achieved global prominence, threatening American producers that might be less competitive (who then seek government intervention and an examination of supply chain risks).

Transformers are core technology in any nation's power grid. Step-up transformers push electricity over long distances to step-down transformers, from which power moves to homes, businesses, and factories. In the last decade, China has become a major manufacturer of power transformers, especially the high-voltage ones, made with hundreds of tons of copper, steel, and aluminum. Most transformers are made outside the United States because they are less expensive than American-made units, and thus meet low-bid requirements of procurement officers. U.S.-China trade conflicts have sensitized administrators to the supply-chain threats transformers pose.

Chinese-made drones also figure in power grid infrastructure. For example, Ohio-based American Electric Power Co. (AEP) owns several drones manufactured by DJI (headquartered in Shenzhen, China, and one of the largest suppliers of commercial drones to the United States). When trade wars started, Department of Homeland Security (DHS) officers suggested that DJI drones could redirect data to Chinese spies, an allegation the company denied. Industry trade group American Fuel and Petrochemical Manufacturers (AFPM) said its companies used DJI drones to "perform inspections, pipeline right of way work, and surveillance" (Sobczak, 2019b) A DHS cybersecurity specialist said the agency had issued industry-wide alerts about the risks associated with using unmanned aircraft system technology produced in China. Drones as well as transformers are on watch lists.

Security concerns prompted the most controversial U.S. actions regarding foreign technology suppliers. In 2017, DHS ordered federal agencies and contractors to cease using software from the Moscow-based cybersecurity firm Kaspersky Lab. Grid regulators (importantly, NERC) then raised alarms over telecommunications equipment from Chinese vendors Huawei Technologies and ZTE. All three foreign firms denied that their products posed national security risks, but agency and trade officials echoed the warnings and those heard earlier at congressional hearings (Sobczak, 2019a).

These concerns about power grid infrastructure paralleled those of the Federal Communications Commission (FCC) as it deliberated how to open up prime space on the 6-gigahertz spectrum band for 5G Wi-Fi competition. (The 5G technology allows data to move at least 100 times faster than the 4G internet platform, and speed is important not only for streaming videos and video games but also for hospitals and other health care facilities.) Yet the FCC's initiative, which seemed open to Huawei, a global 5G leader, ran counter to another national goal of defending critical infrastructure against cyberattacks, and to the effort to restore electricity and water to people as soon as possible after natural disasters (Behr, 2019a).

The United States' developing trade war with China was the background for complaints about foreign suppliers' potential for endangering national security. Huawei has been investigated and sued over allegations that it had stolen intellectual property from Cisco Systems and other competitors, and that it was effectively operating as an arm of the government of China. In addition, a *Wall Street Journal* investigation found evidence that Huawei may have assisted leaders of some African countries in spying on their own citizens (Sobczak, 2019c).

In May 2020, President Trump issued an executive order to ban foreign equipment posing a high security risk from the U.S. power grid. Under the order, the DOE would develop regulations within five months prohibiting purchases of new blacklisted equipment (including operating equipment for natural gas pipelines). DOE also would develop a list of "pre-qualified" vendors and equipment not affected by the order. Given global supply chains and Chinese market dominance in many types of equipment, industry experts questioned whether the ban would work (Behr & Vasquez, 2020). Still, the Trump administration issued a new rule barring Huawei as well as its suppliers from using American technology and software, a further escalation of the trade war (Swanson, 2020).

Catastrophic atmospheric events. There are two kinds of atmospheric events with the potential to significantly disrupt the electricity grid— electromagnetic pulses (EMP) and geomagnetic disturbances (GMD). The main cause of EMP is an atmospheric nuclear explosion sending crippling shock waves. GMD is more associated with solar storms. Both threats are rare, unexpected events, but they may bring on cascading effects, such as damaging current and voltage swings in sensitive electronics. In 2015, Congress gave the energy secretary the sole authority to direct grid defense against a solar storm or electromagnetic pulse attack following a presidentially declared grid emergency (Behr, 2019b). Four years later, Trump gave authority to the DHS to defend the nation against EMP/GMD threats (Behr, 2019c).

The Electric Power Research Institute (EPRI) examined the potential impacts of EMP for three years. It estimated that the explosion of a 1.4 megaton nuclear weapon roughly 250 miles above earth would cripple no more than 14 transformers out of the thousands located across the United States. However, grid relays controlling power flows would suffer substantial damage, causing a widespread blackout. At the end of April 2019, EPRI concluded its study based on laboratory testing and analysis of Los Alamos national lab scenarios. Its findings were that initial pulse impacts could be mitigated in several ways, for example, by shielding cables with proper grounding, using low-voltage surge protection devices, using fiber optics-based communications, and shielding substation control houses among others. Although the combined effects of initial and late pulses from a nuclear explosion in the atmosphere might trigger a regional service interruption, they would not cause a nationwide grid failure. It found that the worst damages would extend to no more than 5–15 percent of all U.S. transformers (Behr, 2019d; Ciampoli, 2019).

A second research study focusing on the threat of solar storms (GMD) was undertaken over a three-year period by the U.S. Geological Survey (USGS).

USGS director Jim Reilly said: "This information will allow utility companies to evaluate the vulnerability of their power grid systems to magnetic storms and take important steps to improve grid resilience" (USGS, 2020).

Extreme weather events accelerated by climate change. From 2003 to 2012, severe weather caused 80 percent of large-scale power outages (and 88 percent of the customers during these weather events were affected by power outages). Whereas early in the period fewer than 80 percent of customers were affected, later in the period, more than 90 percent were (corresponding with the increase in greenhouse gas emissions). In short, outages increased over a decade. During this period, the sources of outages were: storms/severe weather, 59 percent; cold weather/ice storms, 18 percent; hurricanes/tropical storms, 18 percent; tornadoes, 3 percent; and extreme heat/wildfires, 2 percent (Kenward & Raja, 2014).

Of particular interest are forest fires, and the most ruinous of these in 2019 occurred in California. In the previous decade, California had increased its clean energy goals, asking utilities to satisfy most of the electricity needed by using renewables. The state asserted that investing in such a low-carbon future would help counteract climate change effects—such as extremely dry conditions that facilitate forest fires. Ironically, the state's largest owner-invested utility, Pacific Gas & Electric (PG&E), was held responsible for the most disastrous fires that year (because of faulty equipment). A beneficent effect of the terrible 2019 fire season, however, is that California's utilities became experimental grounds for advanced research and deployment of technology such as sensors to deter power line failures that could start fires. Critics of PG&E such as The Utility Reform Network (TURN) say the utility's smart grid expenditures are too little, too late (Behr, 2019e).

Climate change pressures in New York City took a different tack, when an environmental coalition launched a campaign against the city's 16 fossil fuel "peaker" plants (power plants that operate only when there is a high demand for electricity). A new decarbonization law of New York state required that peaker plants' operating permits would be renewed only if they did not violate climate goals. The 5-group Peak Coalition sought to replace the peaker plants with renewable-powered battery storage. To the environmentalists, the peaker plants were "a prime example of environmental racism," as they polluted low-income neighborhoods while collecting billions from their ratepayers (Iaconangelo, 2020).

Responses in these three areas of physical infrastructure, catastrophic atmospheric events, and extreme weather events indicate improvements in assessments, and in a few cases, improvements of technology. The environmental consequences of climate change on America's power grid and energy resources are treated in greater depth in Q33 and Q35.

FURTHER READING

Bakke, Gretchen. *The Grid: The Fraying Wires between Americans and Our Energy Future*. New York: Bloomsbury, 2016.

Behr, Peter. "How 5G high-speed America jolts grid security," *E&E News*, March 21, 2019a.

Behr, Peter. "Trump boosts defense against extreme EMP threat," *E&E News*, March 27, 2019b.

Behr, Peter. "Trump officials to referee fight over EMP grid threat," *E&E News*, April 8, 2019c.

Behr, Peter. "White House prepares to weigh in on hot EMP debate," June 13, 2019d.

Behr, Peter. "Can Calif.'s smart grid double as wildfire defense?" December 11, 2019e.

Behr, Peter and Christian Vasquez. "Trump pulls power grid into U.S.-China battle," *E&E News*, May 4, 2020.

Ciampoli, Paul. "EPRI releases findings from study of high-altitude EMP attack," American Public Power Association, April 30, 2019.

Iaconangelo, David. "Gas peakers to renewable storage: The next climate fight?" *E&E News*, May 11, 2020.

Kenward, Alyson and Urooj Raja. *Blackout: Extreme Weather, Climate Change and Power Outages*. Princeton, NJ: Climate Central, 2014.

Sobczak, Blake. "How hacking threats spurred secret U.S. blacklist," *E&E News*, April 18, 2019a.

Sobczak, Blake. "Feds to energy companies: Beware drones made in China," *E&E News*, May 21, 2019b.

Sobczak, Blake. "No Huawei guarantees," *E&E News*, December 5, 2019c.

Sobczak, Blake and Peter Behr. "China and America's 400-ton electric albatross," *E&E News*, April 25, 2019.

Swanson, Ana. "U.S. delivers another blow to Huawei with new tech restrictions," *New York Times*, May 15, 2020.

U.S. Geological Survey (USGS). "New geoelectric hazard map shows potential vulnerability to high-voltage power grid for two-thirds of the US," March 16, 2020.

Vasquez, Christian. "China's stolen info overload," *E&E News*, February 13, 2020.

Willrich, Mason. *Modernizing America's Electricity Infrastructure*. Cambridge, MA: MIT Press, 2017.

Willson, Miranda. "Fight over 'peaker' plants poses grid climate test," *E&E News*, August 24, 2021.

Q24. CAN THE U.S. POWER GRID BE ADEQUATELY PROTECTED AGAINST RISKS OF ELECTRONIC DISRUPTIONS FROM HACKERS, TERRORIST ORGANIZATIONS, AND HOSTILE NATIONS (E.G., IRAN, RUSSIA, CHINA)?

Answer: The risks can be mitigated but not eliminated. Cybersecurity lapses and failures of the last decades suggest that government efforts in both Republican and Democratic administrations have been seriously deficient to date in keeping up with hackers. Also, issues remain with the lack of clarity in government's emergency response system. When the Colonial Pipeline firm's computer files were hacked in May 2021, company officials did not call DHS's cybersecurity office; they called the FBI. FERC sets mandatory cybersecurity rules for high-voltage electric grids, but the TSA monitoring system relies on companies' voluntary pipeline security checklists (Behr & Vasquez, 2021). Management of the grid often leaves critical safety and security measures to voluntary guidelines.

The Facts: Until the 2015 Russian hacking of three Ukraine power companies, most of the discussion of cybersecurity for America's energy grid was speculative. Because the Russian attack was the first blackout of any country's electrical system by electronic means, it drew immediate attention of U.S. grid security experts, as did a December 2016 attack by suspected Russian hackers on a transmission substation outside Kiev, Ukraine's capital city. The 2015 attack cut power to 225,000 people for six hours; the 2016 attack was repaired within 75 minutes (Sobczak, 2019a).

Resolving potential cyberdefense problems in the United States took a top spot on the electricity grid agenda, but within years private spending by the utilities (as well as governments) declined relative to expenditures for newer problems such as storm hardening, electric vehicle charging, smart grid systems, and renewable energy. When asked at the 2016 midwinter summit of the National Association of Regulatory Utility Commissioners (NARUC) whether power grid facilities critical to U.S. national security were sufficiently protected, Scott Aaronson, a security expert with the Edison Electric Institute, said: "If they were compromised, we would be in a world of hurt." When asked if North American utility operators could, like the Ukrainians, run systems in a "degraded state" so that customers would receive some power during parts of the day, his answer was "sort of" (Behr, 2020).

According to Aaronson, "it would be professional malpractice if we were putting all of our emphasis on 'protect, protect, protect' and not acknowledging that protection—while incredibly important—can't be effective 100% of the time" (Behr & Sobczak, 2016).

Sources of threats. Threats to America's energy infrastructure come from three different sources in increasing degrees of severity. First are individual hackers, foreign or domestic. Second are terrorist organizations. Third are nation-states that are hostile to the United States. It is not in all cases easy to differentiate among the types.

In November 2018, the U.S. Department of Justice (DOJ) indicted eight Russian and Kazakh nationals for conducting millions of dollars' worth of digital advertising fraud, using "botnets" to hack computers. A second indictment alleged that two Iranian citizens spread "SamSam" ransomware through municipal, hospital, and school district networks, causing tens of millions in damages. The previous month, the department had linked Russian government hackers to a cyber espionage effort against Westinghouse Electric (which manufactures nuclear power plants) (Sobczak, 2018b).

In 2018, a Cyber Digital Task Force convened by the DOJ reported that "identity-masking technologies and international investigative barriers pose unique challenges for deterring cyber threats" (DOJ, 2018). Identifying culprits is vitally important, though. Said Katie Nickels, threat intelligence chief for MITRE Corporation's ATT&CK team: "As a defender, do I care if it's North Korea or Russia, or another country? I think at a minimum there is value in tracking it back to a group or campaign" (Sobczak, 2019c). In the context of worsening U.S. relations with Russia in 2022, the Biden administration urged U.S. firms to harden their digital defenses when Russia was "exploring options for potential cyberattacks" (Vasquez, 2022a).

In mid-2019, the *New York Times* reported that American cyber forces had planted malware into Russia's electric power grid that they could use to retaliate if Russia attacked U.S. networks (Nechepurenko, 2019). Security experts predicted that Iran would launch new cyber weapons against U.S. targets to gain advantage in its developing struggles with the United States. Such concerns led Dennis Gilbert, VP and head security officer for Duke Energy Corp., to assert that the power industry was now at "the front lines of a geopolitical struggle" (Behr, 2019).

Nick Rossman, head of IBM's security division, noted that cyberattacks against industrial control systems rose greater than 2,000 percent in 2019 from the previous year. One IBM security report, "X-Force Threat Intelligence Index 2020," attributed part of the increase to nation-state hackers such as the hacker group called Xenotime, which had been linked to a malware attack that resulted in Saudi Arabian energy plant shutdowns (Vasquez, 2020a).

Targets: a) Energy sources (including pipelines and refineries). Before 2018, no cyberattack had physically disrupted the flow of natural gas in the

United States, but pipeline crises made firms and governments aware of the potential for successful hacking. Several areas of concern attracted attention of firms and governments. DHS cited a cyberthreat dating to December 2011, disclosing for the first time that suspected Chinese hackers gained access to the controls of "several U.S. natural gas pipeline companies" (Vasquez & Sobczyk, 2021). The large increase in America's use of natural gas during the 2010s made gas lines feeding from power plants an obvious cybersecurity concern. (The other major energy sources for electricity—coal, nuclear, hydro, and renewables such as wind and solar—used far fewer transmission lines than natural gas.)

Some energy officials and legislators wondered whether oversight of natural gas pipelines by the Transportation Security Administration (TSA) was sufficient. These calls for reform prompted pushback from the American Petroleum Institute (API) and natural gas associations such as the Natural Gas Council (NGC). These energy nongovernmental organizations (NGOs) wanted to maintain the self-regulatory (essentially voluntary) system in effect under the TSA. Representing their views was a new American Petroleum Institute (API) report titled *Defense-in-Depth: Cybersecurity in the Natural Gas and Oil Industry* (API, 2018).

The need for testing and training soon became evident, when a ransomware cyberattack compelled a natural gas company to shut down a pipeline for two days in February 2020. The cause of the attack was a malicious link sent in an e-mail. It first affected the information technology (IT) and then spread into the operating technology (OT) network in a natural gas compression station. DHS's Cybersecurity and Infrastructure Security Agency (CSIS) attributed the breakdown to the lack of proper segregation of the OT from the IT.

The first serious incident of a supply chain cyberattack was on the IT service provider SolarWinds in September 2019. The software was hijacked by hackers suspected to be linked to Russia. Targets included federal agencies, major manufacturing firms, and power utilities. Thousands of SolarWinds' customers downloaded a malicious update containing the Sunburst malware, and hackers moved quickly to set up additional persistent mechanisms to connect to victim networks (Vasquez, 2020b).

The second incident, which took place in May 2021, was an example of a ransomware cyberattack. The suspected firm was Darkside, believed to operate in Russia. In a press release it said its objective was "to make money" and not disrupt society. It obtained $5 million in digital currency from Colonial Pipeline as the firm shut down its 5,500-mile pipeline (which supplied nearly half of the gasoline used on the East Coast) for a week to stop the spread of the attack—a necessity that temporarily pushed gasoline prices dramatically higher. At his first summit with Russian president Vladimir Putin later in 2021, Biden presented a list of 16 critical infrastructure sectors that "should be off limits to attack, period, by cyber or any other means" (Vasquez, 2021). Putin denied playing any role in cybersecurity breaches.

Targets: b) Critical defense institutions. The United States has more than 100 military installations. Base commanders and managers are responsible for steady electricity supply, and they have emergency power generation infrastructure in the event the civilian grid is compromised. The most significant part of the domestic defense establishment is the digital network that serves as America's primary line of defense against cyberattacks from hostile powers.

The Trump administration's pursuit of coal- and nuclear-friendly energy policies extended to efforts to subsidize coal and nuclear power plants near strategic military installations. This directive, however, conflicted with the Department of Defense's "mission assurance" tenet, which emphasizes that it is the responsibility of installation commanders to "identify, design, and install primary power and emergency energy generation systems, infrastructure, and equipment to support their critical energy requirements" (Behr & Northey, 2018). From 2018 to 2022, utility executives said they were not sure whom to listen to regarding providing power to bases (Vasquez, 2022a).

Targets: c) The grid and business supply chains. In 2016, FERC instructed NERC to develop supply chain security standards against cyber threats. This directive came three years after hackers had stolen data from millions of Target customers (by breaking into contractors' networks and Target's accounting system). In 2014, hackers appropriated European industrial control system vendors in order to spy on their customers. Then in 2018, FERC adopted Order 850, requiring large utilities to review business supply chains for points vulnerable to hackers. The commission also directed NERC to protect electronic access and firewalls, with an emphasis on shielding high-risk targets. Among other issues, NERC was instructed to verify that grid operating software had not been tampered with and examine security issues concerning vendors' remote access to critical grid systems (Sobczak, 2018a).

Major cybersecurity agencies. A large number of agencies are involved in U.S. cybersecurity. The Department of Homeland Security (DHS) has two agencies relevant to cybersecurity. The first is the Transportation Security Agency (TSA), which safeguards the security of the traveling public and also the safety and security of the U.S. transportation system. In 2003, after formation of DHS, TSA was absorbed into it. Second is the Cybersecurity Infrastructure Security Agency (CISA), established in 2018.

Since 2009, the Department of Defense (DOD), America's ultimate national security protector, has housed the U.S. Cyber Command.

The Department of Energy (DOE) has two relevant units. The Office of Cybersecurity, Energy Security, and Emergency Response (CESER) focuses on the safety of the electric power grid and oil and natural gas infrastructure. Its Division of Cybersecurity for Energy Delivery Systems (CEDS) specializes on R&D for innovative technologies.

The Department of Transportation (DOT) was established to provide for the safety and security of land surface transportation. In 2004, it created the Pipeline and Hazardous Materials Safety Administration (PHMSA) to oversee pipeline security across the United States.

Finally, the White House includes the office of the Director of National Intelligence (DNI). The position was created after 9/11 (in 2004) to bring unity and order to America's intelligence community. The DNI is head of the 17-member U.S. intelligence community, providing oversight for agencies such as the CIA and FBI.

Interagency coordination. Mid-way through the Trump administration, calls for interagency coordination on cybersecurity increased. A GAO report, noting that "the grid is becoming more vulnerable to attacks," credited DOE and FERC for taking a series of proactive measures to guard the energy sector, but criticized its limited analysis of hacking threats, digital weaknesses, and limitations on quality of data it had used in recent electricity risk studies (Sobczak, 2019b). After a cyberattack on the city of Pensacola, Florida, calls for coordination increased.

Shortly after taking office in January 2021, President Biden remarked on the need to counter cyberthreats to the American power grid. In April 2021, the departments of Energy and Homeland Security announced a 100-day plan to secure the grid from hackers. The plan encouraged operators and grid owners to improve detection, mitigation, and forensic capabilities. It did not, however, continue a policy implemented by the Trump administration of blocking foreign-made grid equipment from being installed in U.S. networks (Vasquez, 2020c). In May 2021, Biden signed an executive order to boost federal government defenses against cyberespionage campaigns. The order required more rigorous standards for federal contractors and a cyber review board to analyze lessons learned from major breaches (Vasquez, 2021). CISA expanded its security partnerships with Honeywell, Bechtel, GE, and other industrial technology companies (Vasquez, 2022c). The FERC also planned new orders in 2022 to compensate utilities for addressing cybersecurity threats going beyond what federal regulators require (Willson, 2021).

At her Senate confirmation hearing, Biden's secretary of energy, Jennifer Granholm, noted that "cybersecurity threats to the energy sector continue to grow as increasing segments and components of energy systems are interconnected and managed remotely (*E&E News*, 2021). She spoke of both IT systems and operational technology (OT) that controls large networks of power (or pipe) lines, wires, and nodes. With its emphasis on renewable energy, the Biden administration has also emphasized the importance of securing electric vehicle chargers, wind technology, rooftop solar panels, and other green energy options from cyber threats. Amid escalating campaigns of misinformation from multiple sources, global cyberthreat specialist Howard Marshall said, "There's no playbook for this" (Vasquez, 2022d).

Q25 begins a new chapter in this volume, on legal and regulatory frameworks.

FURTHER READING

American Petroleum Institute (API). "Defense-in-depth: Cybersecurity in the natural gas and oil industry." Washington, DC: 2018. https://www.api.org/news-policy-and-issues/cybersecurity/defense-in-depth-cybersecurity-in-the-natural-gas-and-oil-industry

Behr, Peter. "More utilities share a foxhole on cyber front," *E&E News*, June 18, 2019.

Behr, Peter. "Grid's 'big problem' far from fixed," *E&E News*, February 12, 2020.

Behr, Peter and Blake Sobczak. "Utilities look back to the future for hands on performance," *E&E News*, July 21, 2016.

Behr, Peter and Hannah Northey. "Do DOE and the Pentagon march together on grid security?" *E&E News*, October 18, 2018.

Behr, Peter and Christian Vasquez. "Pipeline hack exposes federal holes in U.S. cyber oversight," *E&E News*, May 14, 2021.

Dragos. *2019 Year in Review: ICS Vulnerabilities*, 2020.

Fidler, David. "Year in review: Cyber threats and the mid-term U.S. elections," Council on Foreign Relations. https://cfr.org/blog/year-review-cyber-threats-and-mid-term-us-elections

Matishak, Martin. "Intelligence heads warn of more aggressive election meddling in 2020," *Politico*, January 29, 2019.

Nechepurenko, Ivan. "Kremlin warns of cyberwar after report of U.S. hacking into Russian power grid," *New York Times*, June 17, 2019.

Sobczak, Blake. "FERC signs off on cyber rules for 'highest-risk' grid systems," *E&E News*, October 18, 2018a.

Sobczak, Blake. "Rod Rosenstein to cyber criminals: 'We can catch you,'" *E&E News*, November 30, 2018b.

Sobczak, Blake. "The inside story of the world's most dangerous malware," *E&E News*, March 7, 2019a.

Sobczak, Blake. "Battle lines form over pipeline cyberthreat," *E&E News*, July 25, 2019b.

Sobczak, Blake. "Not over yet CrashOverride," *E&E News*, CyberSecurity update, September 19, 2019c.

Sobczak, Blake. "Watchdog warns FERC, DOE of 'significant' hacking risks," *E&E News*, September 25, 2019d.

U.S. Department of Justice. *Report of the Attorney General's Cyber Digital Task Force.* Washington, DC: 2018. https://www.justice.gov/cyberreport

Vasquez, Christian. "Cyberattacks on industrial systems spike 2,000%," *E&E News*, February 11, 2020a.

Vasquez, Christian. "Cyberattack shut down gas pipeline for days—DHS," *E&E News*, February 19, 2020b.

Vasquez, Christian. "NSA issues rare warning as Russian hackers threaten energy," *E&E News*, July 26, 2020c.

Vasquez, Christian. "Ransomware is disrupting energy. It might get worse," *E&E News*, June 17, 2021.

Vasquez, Christian. "'Evolving intelligence' puts U.S. energy industry on high alert," *Energywire*, March 22, 2022a.

Vasquez, Christian. "DOE, Pentagon misaligned on military grid defense program—report," *Energywire*, March 23, 2022b.

Vasquez, Christian. "U.S. cyber agency expands alliance with energy giants," *Energywire*, April 21, 2022c.

Vasquez, Christian. "War, fear, 'hactivist' zeal are upending energy cybersecurity," *Energywire*, April 4, 2022d.

Vasquez, Christian and Blake Sobczak. "China hacking threat prompts rare U.S. pipeline warning," *E&E News*, July 21, 2021.

Willson, Miranda. "FERC to issue grid, cybersecurity, hydro rules," *E&E News*, December 14, 2021.

5

Legal and Regulatory Frameworks

The overall framework of government and politics in the United States is authorized by the U.S. Constitution, its 27 amendments, and Supreme Court interpretations of those documents that have been handed down over the course of American history.

Laws are decisions of legislative bodies (at federal or state levels) implemented and enforced by regulatory and law enforcement institutions.

Regulations are the products of local, state, and federal regulatory agencies charged with carrying out the laws passed by lawmakers. Regulations (and rules) have the force of law (Kerwin, 1999).

Rise of national power. One hundred years ago, the most prominent political leaders were governors and state legislators, but their influence sometimes paled in comparison to captains of industry such as John D. Rockefeller and Henry Ford. Then two events shook the economies and societies of the world—the Great Depression of the 1930s and World War II, which raged from 1939 to 1945. In the first five years of the Depression, the American GDP (gross domestic product) fell by 50 percent, rates of unemployment reached 25 percent, and homelessness skyrocketed.

These dire conditions gave President Franklin Delano Roosevelt, who entered the White House in March 1933 with an expansive view of presidential power and broad public support, the opportunity to lay down what became the foundation of the modern U.S. welfare state. He created a nationwide social insurance program with passage of the Social Security Act of 1935. In addition, his administration launched ambitious jobs programs for youth and laid-off workers, as well as a range of farm relief programs. What Roosevelt called a "new deal" of activist government programs won support at the polls. Full economic recovery in America did not occur, however, until war broke out in Europe and Asia. The U.S. war effort galvanized the nation's industrial might and fully employed workers, and America emerged from World War II with a healthy and dynamic economy (Heilbroner & Milberg, 2017).

In the era of the 1930s and 1940s, regulatory authority of the U.S. government had expanded in response to the Depression. In the postwar era, the U.S. Constitution, once applied primarily to relations between federal and state governments, was applied to individual citizens directly. Racially discriminatory provisions of state constitutions were overridden by the U.S. Supreme Court, ending school segregation and Jim Crow laws of the South, as well as protecting individuals (such as those accused of lawbreaking) against the state. By the 1970s, the regulatory reach of the U.S. government into the economy and society had extended deeper than ever before.

During this era of increased regulation, new laws were crafted to protect the environment from pollution; secure the civil rights of African Americans, women, and other minorities; and address other perceived societal ills. But these new regulations also prompted a backlash from critics who decried them as burdensome to American businesses and oppressive to individual liberties. In 1980, Republican Ronald Reagan rode this growing anxiety about excessive regulation to the White House, where he tried to roll back many regulations as part of an effort to "get government off the backs of the American people." Since that time, American politics has become increasingly entangled in debate about where sensible regulation ends and intrusive government overreach begins.

Chapter outline. Six questions organize discussion in this chapter. Questions 25 and 26 address the separation of powers (SOP) among the three branches of government and the increased politicization of agency leadership and actions.

The next set of three questions (27 through 29) focus on regulatory agencies and the 2 million members of the federal civil service. The main topic of Q27 is whether scientific research—including research related to energy and associated environmental issues—has become politicized. Q28 asks whether the Trump administration proposal to "modernize" the National Environmental Policy Act (NEPA) harkened back to previous reform attempts. Q29 raises a question asked by both Republicans and Democrats when out of power: Have regulatory agencies and bureaus been "captured" by the interests and industries they are supposed to regulate? Finally, Q30 explores whether federal–state conflicts over energy policies and environmental regulation have widened in recent years.

FURTHER READING

Eisner, Marc Allan. *The State in the American Political Economy*, 2nd ed. New York: Routledge, 2014.

Heilbroner, Robert and W. Milberg. *The Making of Economic Society*, 17th ed. New York: Pearson, 2017.

Kerwin, Cornelius M. *Rulemaking: How Government Agencies Write Law and Make Policy*, 2nd ed. Washington, DC: Congressional Quarterly Press, 1999.

Q25. HAS DIVIDED CONTROL OF THE EXECUTIVE AND LEGISLATIVE BRANCHES CONSTRAINED PRESIDENTIAL ADMINISTRATIONS FROM MAKING MAJOR CHANGES IN U.S. ENERGY POLICY IN RECENT YEARS?

Answer: Yes.

The Facts: The framers of the U.S. Constitution used the term separation of powers (SOP) to describe the relationship of institutions. Today, many political scientists define SOP as "separated institutions sharing and competing for power." James Madison famously established the principle at work in *The Federalist* #51, where he argued that concentrated powers were the enemy of liberty and could be avoided by giving those administering each branch the "necessary constitutional means, and personal motives, to resist encroachments of the others," to wit: "Ambition must be made to counteract ambition. The interest of the man must be connected with the constitutional rights of the place" (Rutland et al., 1977).

Initially in American history the president competed with Congress at a disadvantage. The framers named the First Article of the Constitution for the Congress, believing it to be closest to the people. (The House had been directly elected by the people since 1789, but the Senate was indirectly elected [appointed by the upper house of the state legislatures] until passage of the Seventeenth Amendment in 1913.) Framers thought the Supreme Court (the only court mentioned in the Constitution) would be the weakest branch, because it lacked either the power of the "purse" (the power of appropriation, held by Congress) or the "sword" (the power to wage war and command the nation's military, which resided with the office of the president).

Trump administration regulatory rollback efforts. A number of Trump administration proposals to make regulatory changes to environmental and energy rules were vulnerable to court reversals. In some cases, the courts stepped in to halt regulatory rollbacks because the Trump administration did not provide adequate analyses supporting a repeal or replacement of Obama-era rules. In other cases, judges held that Trump officials did not take sufficient consideration of benefits lost from rolling back a regulation (Belton & Graham, 2019).

Toward the end of President Trump's term, however, several court rulings were more favorable. In a major Superfund case (Atlantic Richfield Co. v. Christian), for example, the Supreme Court sided with the owner of the oldest and largest cleanup site in the nation in a dispute over remediation. Trump's Department of Justice had supported the legal position of the cleanup site owner (Hijazi, 2020).

Perhaps the clearest evidence of the courts' continued ability to restrain the executive were decisions in high-stakes environmental issues. For instance, the majority (6-3) opinion in *County of Maui v. Hawaii Wildlife Fund* steered between a major expansion of the Clean Water Act and allowing regulated entities (such as wastewater plants) to avoid permitting requirements (King, 2020). From 2017 through 2020, the Trump administration deregulated elements of the Obama administration energy and environmental policy legacy: the Clean Power Plan (to reduce U.S. carbon dioxide emissions by more stringent implementation of the Clean Air Act), and the Waters of the U.S. (WOTUS) rule (defining which waterways and wetlands are automatically protected under the Clean Water Act). Court rulings on these changes, however, effectively deferred Trump administration rollbacks for consideration by the Biden administration (see Q31 and Q35).

The Constitution gives Congress primary budget authority, which it exercises through its powers of authorization and appropriation. Since the Great Depression, presidents have whittled away at this authority, and Trump was no exception. In 2017, he proposed and the Republican majority in Congress adopted the Tax Cuts and Jobs Act, which significantly lowered tax rates for corporations and high-income Americans. (The Congressional Budget Office [CBO] estimated it increased budget deficits by $2.3 trillion over the 2018–28 period [Hall, 2019].) Simultaneously, Trump sought to reduce government spending in many nonmilitary areas. For example, he proposed taking $2 billion in funding from the Energy and Interior departments and the Environmental Protection Agency. Congress, however, restored funding of these agencies and departments to status quo levels. This ensured that programs in which members were invested (such as cleanup of toxic waste sites under Superfund) and government jobs would not be cut (Bogardus, 2019).

In its deliberations on the FY 2020 budget (beginning on October 1, 2019), members of appropriations committees in the House and Senate rejected deep cuts proposed by the Trump administration for several EPA, Interior, and Energy programs. One issue was the White House proposal to move the headquarters of the Bureau of Land Management (BLM) from Washington, DC, to Grand Junction, Colorado. The rationale provided was that most U.S. public lands under BLM management were in the Western states, but Congress did not approve funding for the move (Lumney & Bogardus, 2019). When the Biden administration took over in 2021, new Interior secretary Deb Haaland canceled the move, leaving Grand Junction as the BLM's regional hub in the West.

Most presidents decline to influence their successor's term with "midnight" orders, but on January 18, 2021 (two days before the inauguration of Democrat Joe Biden, who defeated Trump in the November 2020 presidential election), President Trump issued two executive orders relevant to energy policy. His first executive order banned career staff in federal agencies from endorsing new regulations, while the second order was justified as an attempt to "protect Americans from overcriminalization through regulatory reform" (Brugger, 2021). This too was an attempt to dilute enforcement of environmental laws and regulations. Both of these executive orders were reversed by the Biden administration.

Court restrictions on Biden executive actions. More serious obstacles to the Biden administration actions came from the courts and Congress.

The incoming Biden administration immediately froze litigation pushed by the previous administration in support of environmental rule rollbacks. Instead, Biden issued an executive order on his first day in office instructing agency heads to carefully review those rules to ascertain whether they should be reversed (King, 2021a).

At the Supreme Court, however, legal experts cautioned the Biden administration about potential changes in Supreme Court procedures that might influence interpretation of environmental law.

The procedure is called "standing doctrine," by which a court may dismiss a case if the plaintiff has not demonstrated concrete harm or suffering meeting eligibility criteria for judicial review (under the judiciary article of the Constitution). Some legal experts believe that the conservative majority of the Supreme Court will narrow its definition of "standing" to make it harder for people, communities, and organizations to file lawsuits on the basis of harms suffered from climate change, industrial pollution, and habitat loss. Several justices, including John Roberts and Amy Comey Barrett, have taken narrower views of standing that may exclude subnational governments (states, local governments) and environmental nongovernmental organizations (NGOs) (King, 2021b).

Another emerging threat to the Biden (or other center-left) administration is the Supreme Court's "shadow docket"—unsigned rulings by the Court that hand down legally binding decisions without any explanation of their legal basis. For example, in 2015, the Obama EPA adopted the Clean Power Plan, a rule curbing carbon dioxide emissions from coal-fired power plants. The Supreme Court stayed the rule from taking effect, without hearing arguments or issuing a formal opinion.

The Trump administration benefitted more than it lost from the shadow docket. His administration won 28 of 41 cases brought via the shadow docket, whereas the Bush and Obama administrations filed fewer such cases. Two August 2021 cases brought by the Biden administration were rejected by the Supreme Court using this method. It ended the pandemic-related federal moratorium on residential evictions, explaining the decision in a short opinion. Then the high court rejected the Biden administration's request to rescind the Trump immigration policy forcing thousands of asylum seekers to "remain in Mexico" awaiting U.S. hearings. This decision was not transparent; the explanation was in an order of two paragraphs. It seems unlikely that the Biden administration will turn to the Supreme Court often for emergency/shadow docket rulings (Hurley, 2021).

FURTHER READING

Belton, Keith and John Graham. *Trump's Deregulatory Record: An Assessment at the Two-Year Mark.* Center for Policy Research, American Council for Capital Formation, 2019. https://policyinstitute.iu.edu/doc/mpi/accf-white-paper.pdf

Bogardus, Kevin. "Congress bucks White House on environmental spending," *E&E News*, February 14, 2019.

Bogardus, Kevin. "Biden's EPA noms face sluggish confirmation," *E&E News*, August 19, 2021.

Brugger, Kelsey. "Trump issues new executive orders limiting regulations," *E&E News*, January 19, 2021.

Cahlink, George. "Energy aides named in articles of impeachment against Trump," *E&E News*, December 10, 2019.

Hall, Keith. "How the 2017 tax act has affected CBO's GDP and budget projections since January 2017," February 28, 2019. https://www.cbo.gov/publication/54994

Hijazi, Jennifer. "DOJ celebrates legal wins for Trump's agenda," *E&E News*, April 23, 2020.

Hurley, Lawrence. "Analysis: Biden's Supreme Court losses prompt more 'shadow docket' scrutiny," *Reuters*, August 27, 2021.

King, Pamela. "Roberts' court finds the middle in high-stakes enviro term," *E&E News*, July 13, 2020.

King, Pamela. "Courts freeze Trump rule litigation," *E&E News*, February 9, 2021a.

King, Pamela. "Inside a legal doctrine that could silence enviros in court," *E&E News*, July 19, 2021b.

Lumney, Kellie and Kevin Bogardus. "Senate unveils Interior-EPA bill, rejects money for BLM move," *E&E News*, September 24, 2019.

Richards, Heather. "Nada Culver shakes up BLM, oil and climate policy," *E&E News*, April 7, 2021.

Rutland, Robert, Charles Hobson, William Rachal, and Frederika Teute. *The Papers of James Madison, vol. 10.* Chicago: University of Chicago Press, 1977.

Q26. ARE PARTISAN POLITICAL CONSIDERATIONS INCREASINGLY DICTATING HOW FEDERAL AGENCIES AND DEPARTMENTS—INCLUDING THOSE RELATED TO THE ENERGY INDUSTRY— OPERATE AND WHO LEADS THEM?

Answer: Yes.

The Facts: *Appointing loyalists.* While the number of political appointments to the Trump administration was not significantly larger than in previous administrations, government experience was less important for candidacies than public criticism of the previous administration's policies and demonstrated loyalty to the Trump campaign. These priorities were on display with a number of cabinet appointments relevant to energy policymaking, including Trump's appointment of Scott Pruitt to head the Environmental Protection Administration (EPA), Rick Perry to run the Department of Energy (DOE), and Ryan Zinke to be secretary of the Department of the Interior (DOI). All had been critical of Obama administration positions on climate change and were strong defenders of fossil fuel industries (Davenport, 2016). Scandals forced both Pruitt and Zinke from office, while Perry resigned at the end of 2019.

The Trump administration also prioritized conservative ideology and loyalty to the president in filling other positions. In the Interior Department, acting BLM director William Pendley was openly scornful of public employees, saying "the deep state is alive," and described Democratic candidates who sought to halt oil and gas leasing on federal land as "absolutely insane" (Yachnin, 2019). The administration also appointed Alex Fitzsimmons as acting deputy assistant secretary for energy efficiency. Fitzsimmons previously had worked for Fueling U.S. Forward, a nonprofit set up by Koch Industries to build support for fossil fuels among minorities and low-income people. In 2014, he dismissed wind energy as "expensive" and "unreliable." However, when he later served as chief of staff at DOE's Office of Energy Efficiency and Renewable Energy (EERE), he reversed himself, saying wind was "affordable and reliable."

Purging enemies. After the first impeachment of President Trump for his efforts to pressure Ukrainian president Zelensky to interfere in America's 2020 presidential election, loyalty to the president became an even greater point of emphasis in many public agencies and departments. After the Senate declined to convict the president, his new director of the Office of Presidential Personnel, Johnny McEntee, ordered a freeze on hiring and reportedly instructed departments to find those not loyal to the president so they could be removed. Scholar Paul Light called it "the biggest assault on the nation's civil service system since the 1883 Pendleton Act ended the spoils system." Kathryn Tempas, a Brookings scholar, produced data showing that turnover among senior staff hit 82 percent in just three years, greater than for any of the five previous presidents in their first four years (Baker, 2020). The purging even extended to inspectors general (IG) in different agencies who were perceived to be insufficiently loyal to Trump.

In late October 2020, Trump issued an executive order establishing "Schedule F." This new classification in the civil service was for employees deemed substantially involved in policymaking, who could be fired peremptorily. EPA identified 579 positions that fit this new classification in air, water, and toxic chemicals regulation, as well as research and administration. The president's executive order stated: "Agencies need the flexibility to expeditiously remove poorly performing employees without facing extensive delays or litigation" (Bogardus, 2021). The order gave agencies 90 days to complete the review (one day before inauguration of the new president). The order was among the first rescinded by the Biden administration.

Leaving positions vacant while interim officials do the president's work. Federal legislation specifies which executive positions required U.S. Senate confirmation. For example, the director of an interior department bureau required confirmation, but a deputy director did not. This option, for example, enabled the Trump administration to appoint Karen Budd-Falen, a well-known property rights lawyer, to become deputy solicitor for Fish, Wildlife and Parks over the objections of Democrats who had blocked other conservative nominees for Interior Department posts (Streater & Doyle, 2018).

Reorganizing agencies/departments. A major initiative of Ryan Zinke as Interior Department secretary was to reorganize the agency's nearly 70,000 employees. Initially, the design was to have 12 new unified regions that aligned the several bureaus of DOI along state boundaries and watersheds (Yachnin, 2018). Governors of Colorado, Nevada, and New Mexico (sites of proposed "unified regions") supported the proposal, as did both Republican and Democratic senators. The agency most impacted was BLM, whose proposed new headquarters would be in Grand Junction, Colorado. The coalition of reorganization supporters claimed that most of DOI's work was in the Western states, and the move would save money.

By the outset of the Biden administration, 41 of 328 reassigned BLM employees had moved West; most retired or took other jobs in the Washington, DC, area. In the last nine months of the term, the Trump administration replaced departed employees with new hires who wanted to be in Grand Junction. In an unanticipated twist, Colorado's two Democratic senators asked Biden to keep the new BLM office in Grand Junction open and beef it up (Streater, 2021a). New Interior secretary Deb Haaland subsequently announced that the Biden administration intended to return BLM headquarters to Washington, DC, but leave a Western regional office in Grand Junction (Streater, 2021b).

FURTHER READING

Baker, Peter. "Trump's efforts to remove the disloyal heightens unease across his administration," *New York Times*, February 22, 2020.

Bogardus, Kevin. "How EPA complied with Trump order to dismantle 'deep state,'" *E&E News*, January 25, 2021.

Brugger, Kelsey. "Greens' complaint charges Trump admin with purging enemies," *E&E News*, March 5, 2020.

Brugger, Kelsey and Hannah Northey. "Ex-Hill aide, vocal Trump ally returns to EPA in key role," *E&E News*, March 16, 2020.

Davenport, Coral. "Trump is Said to Offer Interior Job to Ryan Zinke, Montana Lawmaker," *New York Times*, December 13, 2016.

Streater, Scott. "Bernhardt keeps Pendley atop BLM as lawsuit looms," *E&E News*, May 5, 2020.

Streater, Scott. "With 287 staffers gone, should Biden return BLM to D.C.?" *E&E News*, January 29, 2021a.

Streater, Scott. "3 big questions surround BLM headquarters move back to D.C.," *E&E News*, September 20, 2021b.

Streater, Scott and Michael Doyle, "Controversial lawyer named to key fish and wildlife post," *E&E News*, October 15, 2018.

Yachnin, Jennifer. "Zinke's reorganization push will likely go on," *E&E News*, December 17, 2018.

Yachnin, Jennifer. "BLM chief scorned 'deep state,' 'Pocahontas,' ESA 'hammer,'" *E&E News*, October 17, 2019.

Q27. DID THE TRUMP ADMINISTRATION POLITICIZE SCIENTIFIC RESEARCH IN U.S. ENERGY AND ENVIRONMENTAL POLICYMAKING?

Answer: Yes.

The Facts: Scientific research conducted by or on behalf of government agencies and departments has long been an important element in policymaking, in energy and many other realms. In the twenty-first century, however, conservative officials and lawmakers in particular have been accused of interfering in that research or muzzling findings that do not support their own political ideologies or allies.

Definitions and general context. Motivating the animus against traditional science was government regulation of health and industry starting in the 1970s and 1980s. Citizens for a Sound Economy (CSE), a lobbying group founded by the Koch brothers (owners of the second-largest private enterprise in the United States and major donors to libertarian and conservative causes), warned that new air quality standards would force factories to close. CSE leader C. Boyden Gray (heir to the Reynolds tobacco fortune and GOP activist) developed an air quality standards coalition coordinated by the National Association of Manufacturers to oppose limits on air pollution. In 1997, CSE organized demonstrations at hearings and a publicity campaign attacking scientific studies (including studies by Harvard scientists and the American Cancer Society). When attempts in Congress to establish EPA's authority to set standards seemed likely to bear fruit, a tobacco and oil industry consultant named Steve Milloy launched the "Stop Secret Science" media blitz, and the anti-"secret science" movement was born (American Independence Institute, 2014).

The concept "secret science" was coined to discredit well-founded scientific research that follows traditional practices of observation, experimentation, and replication of results. The modifier "secret" used to undercut public trust in scientists and their research efforts refers to basic data that is kept confidential to protect the privacy of individuals or for other legitimate reasons.

Science policy in the Obama administration. Republican congressman Lamar Smith of Texas held the anti-"science" banner aloft when he chaired the House Science, Space and Technology Committee after the 2014 midterm elections. One of the consistent themes in his unbroken (16 terms) record in the House was as a climate change denier, and he often used his chairmanship to conduct battle with the Environmental Protection Agency (EPA). He crafted the Secret Science Reform Act, a two-paragraph bill that would have stopped the EPA from taking any action, whether regulatory or enforcement, unless it were based on scientific research containing no confidential data. Had Smith's bill been enacted, it would have halted all public health and air quality regulation, both of which rely heavily on confidential research data. The legislation failed to pass, however, and Smith resigned from the House in 2016 for a career as a lobbyist.

President Obama's strong belief in science-based decision-making led to his creation of a White House Office of Science & Technology Policy (OSTP) with a 100-person staff. An early Obama administration focus was a Clean Power Plan to combat climate warming, to be spearheaded by the EPA. The plan's call for dramatic cuts in carbon pollution from U.S. power plants (the largest source of air pollution) stimulated a storm of criticism from the oil, gas, and coal industries. During the 2016 presidential election campaign, Donald Trump called climate change a hoax and successfully derailed the Clean Power Plan, at least as of early 2022.

Changes of science policy in the Trump administration. In March 2018, EPA administrator Scott Pruitt signed a rule proposal for the agency that reflected President Trump's attitude toward scientific studies and research that did not dovetail with his policy preferences. The new rule would require scientific research by the EPA to "be transparent, reproducible and able to be analyzed by those in the marketplace," unlike the existing "secret science." Critics pointed out that the proposed policy would violate several laws EPA was obligated to observe in order to protect water, air, and land from pollution.

Pruitt's successor as director, Andrew Wheeler, expressed support for the proposed rule, as did many Senate Republicans. Senator Mike Rounds of South Dakota, for example, charged that "lack of transparency at the EPA had led the agency to seek out the science that supports a predetermined policy outcome." Opponents thought the proposed rule fit the broader pattern of "regulatory capture" by industries the agency was charged with overseeing (Joselow, 2018; see also Q28). ("Regulatory capture" is a term used to describe the belief that regulatory agencies such as the EPA sometimes come under undue influence of the industries and other interests that they are supposed to regulate [e.g., oil and gas companies] and no longer operate in the public interest.)

The Union of Concerned Scientists, a moderately liberal association of U.S. scientists, issued a report called "The State of Science in the Trump Era" (Carter et al., 2019). It complained about inappropriate corporate influence within Trump's EPA and charged the Trump administration with having "significantly undermined" the role of science in government decision-making. For example, the report castigated the Interior Department for ending independent studies of the impacts of oil, gas, and mining industries on public safety and health.

Soon thereafter, the proposed rule (by then called "Strengthening Transparency in Regulatory Science") was leaked to the *New York Times*. Critics alleged that the proposed rule aimed to weaken or eliminate air pollution regulations (by stringently limiting use of human health studies), which industry found cumbersome. Said a persistent critic, Michael Halpern (deputy director of the Center for Science and Democracy, Union of Concerned Scientists [UCS]): "This is not a scientific debate, . . . (It is) a political debate about whether EPA is going to be able to consider the best possible evidence." Only a few academics, such as Susan Dudley, director of the George Washington University Regulatory Studies Center, supported the change: "Greater transparency could encourage more openness and constructive discussion about science and policy" (Brugger, 2019a). Ultimately, the rule was eliminated by the Biden administration in May 2021.

The Trump administration tweaked its science policy in two other ways. First, it employed a rarely used act of Congress called the Information Quality Act (IQA) to require agencies to specify their definition of "influential information" and make it publicly available.

Second, after several years of inactivity, the Trump administration revitalized the Office of Science and Technology Policy by creating a new panel, the President's Council of Advisors on Science and Technology. Initially, six of the new members were corporate executives—from IBM, HP, Bank of America, among others, and included a Dow Chemical VP whose firm contributed $1 million to the president's inauguration (Brugger, 2019b).

Scientific advisory committees during the Trump years. Advice from science advisory committees helps drive government decisions that have saved lives, improved public health, and protected the environment. For example, such committees carried out analysis that led to the phaseout of neurologically damaging lead in paint and gasoline in the 1970s; the establishment of air quality standards to guard against pollutants; and increased regulation of chemicals and metals damaging to air, water, and soil quality (UCS, 2019).

The Trump administration's populist and nationalist goals (strongly supported by the president's base) often ran afoul of the findings of scientific advisory committees. As a result, administration officials increasingly challenged individuals' qualifications for service, composition of committees, and committee missions and mandates.

Traditionally, members of advisory bodies are selected based on their expertise, for example, for scientists, the number of peer-reviewed papers published in high-impact journals in their areas of specialization. Too, panels offering advice to officials typically aspire to represent a balance of perspectives and be free of conflicts of interest. In these areas, a 2019 General Accounting Office report found high-profile EPA advisory committees in the Trump administration to be deficient. It questioned the qualifications of members that the Trump administration had invited to join the committees, found that academic scientists were underrepresented, and warned that investigations of potential conflicts of interest of members had not been made (GAO, 2019).

The number of advisory committees also came under attack from the Trump administration. There are approximately 1,000 advisory committees in federal agencies, with more than 60,000 members; most committees operate under the Federal Advisory Committee Act (FACA). In 2019, President Trump's executive order directed agencies to cut their advisory boards by at least one-third within three months (an earlier order directed agencies to repeal two regulations for every new one created). Former EPA officials feared that the executive order would adversely affect its scientific work; advocates and observers thought that elimination of advisory boards would lessen outside scrutiny of the agency (Green & Beitsch, 2019).

Another issue was committee structure. In EPA, the Scientific Advisory Board (SAB) was the overarching frame for the agency's 22 committees, but its 44-member board was too large for it to make decisions easily. In some cases, missions of individual committees were specifically mentioned in authorizing acts of Congress; but in most, the agency director or department secretary determined the committee's responsibilities. The Trump administration's criticisms in this regard were legitimate, but critics claimed that the administration's proposed "solutions" were arbitrary and antagonistic to the role of science in modern American society.

Biden administration actions on scientific research. At the request of the incoming Biden administration, Montana District Court Judge Morris returned the "Strengthening Transparency in Pivotal Science Underlying Significant Regulatory Actions and Influential Scientific Information" (aka "open science") rule of the Trump administration to the EPA. The federal judge found that the rule, which had been finalized late in the Trump administration, had not been properly implemented by EPA (King, 2021).

A month later, the acting secretary of the Interior Department rescinded the order signed by Bernhardt in 2018. The rescinded rule was similar to that of EPA, except in the case of Interior it precluded the department from using the best (and sensitive) information "regarding sacred sites or rare and threatened species to inform complex policy decisions." An assistant secretary for water and science, Tanya Trujilo, commented: "Today's Order puts the evaluation and decision-making authority regarding scientific information back where it should be, in the hands of scientists" (Doyle, 2021).

By May 2021, the EPA had put the final touches on a regulation vacating the final version of the Trump administration's rule. Paul Billings, an official of the American Lung Association, hailed the Biden administration's decision to scrap what he called the Trump administration's "censoring science" rule: "The EPA vacating this rule and putting an end to this effort to cherry-pick scientific studies and undermine the health benefits for curbing air pollution should be thrown on the scrap heap of terrible Trump-era regulations" (Brugger, 2021a).

Biden administration reconstitutes advisory committees. Shortly after his confirmation as EPA administrator, Michael Regan fired all members of two scientific advisory committees—the Science Advisory Board (SAB) and Clean Air Scientific Advisory Committee (CASAC)—saying a "reset" was needed. Regan's move was a response to a Trump-era decision to ban all EPA grant recipients from serving on advisory panels and to decline to reappoint members to new terms. Trump officials claimed that these rules were implemented to keep panels free of undue agency influence, but critics alleged that it was a transparent ploy to give industry more power on the committees and weaken the role of academics and scientists in setting environmental policy (Reilly, 2021).

EPA administrator Regan addressed EPA employees by promising to "restore the role of science and transparency" at the agency, but his decision to remove all members of the two bodies was criticized as extreme in some quarters. Former SAB chair Michael Honeycutt said, "I don't think two wrongs make a right. . . . Pruitt and Wheeler were wrong to exclude scientists who receive EPA funding from serving on the board, but . . . Regan was also wrong to fire the board." This led the *E&E News* reporter Kevin Bogardus to wonder whether this was the onset of "a scorched-earth policy that after every presidential election cycle could potentially undermine the agency's core mission to evaluate science and regulate?" (Bogardus, 2021). The reset did increase tensions between industry and environmental groups, but also produced a new cohort of distinguished science advisers including two Nobel laureates, five MacArthur "genius" fellows, two former cabinet secretaries, and many members of National Academies (Brugger, 2021b).

FURTHER READING

American Independence Institute. "Republicans wage anti-'secret science' campaign against the EPA," *Huffington Post*, June 25, 2014. https://www.huffpost.com/entry/secret-science-epa_n_5529521

Bogardus, Kevin. "Regan's contentious bid to reset EPA scientific integrity," *E&E News*, April 27, 2021.

Brugger, Kelsey. "Trump admin advances high-impact 'secret science' rule," *E&E News*, November 12, 2019a.

Brugger, Kelsey. "Trump adds academics to industry-stacked science panel," *E&E News*, November 13, 2019b.

Brugger, Kelsey. "Trump admin expands reach of secret science proposal," *E&E News*, March 4, 2020.

Brugger, Kelsey. "Final ax coming for contentious Trump transparency rule," *E&E News*, May 21, 2021a.

Brugger, Kelsey. "Biden announces new cadre of outside science advisers," *E&E News*, September 22, 2021b.

Carter, Jacob, Emily Berman, Anita Desikan, Charise Johnson, and Gretchen Goldman. *The State of Science in the Trump Era: Damage Done, Lessons Learned, and a Path to Progress*. Center for Science and Democracy at the Union of Concerned Sciences, January 2019. https://www.ucsusa.org/resources/state-science-trump-era

Doyle, Michael. "Biden admin ditches Trump 'open science' order," *E&E News*, March 3, 2021.

Green, Miranda and Rebecca Beitsch. "Trump directs agencies to cut advisory boards by 'at least' one third," *The Hill*, June 14, 2019.

Joselow, Maxine. "Wheeler punts Pruitt's science overhaul," *E&E News*, October 17, 2018.

King, Pamela. "Judge scraps Trump's 'secret science' rule," *E&E News*, February 1, 2021.

Reilly, Sean. "Decks cleared: Regan fires science board members," *E&E News*, March 31, 2021.

Union of Concerned Scientists (UCS). *Scientific Advisory Committees*. June 13, 2019. https://www.ucsusa.org/resources/scientific-advisory-committees

U.S. Congress, Governmental Accountability Office (GAO). *EPA Advisory Committees: Improvements Needed for the Member Appointment Process*, 2019. https://www.gao.gov/products/GAO-19-280

U.S. Environmental Protection Administration. *Fact Sheet: Overview of the Clean Power Plan*. Washington, DC, 2015. https://19january2017snapshot.epa.gov/cleanpowerplan/fact-sheet-overview-clean-power-plan_.html

Q28. DID REGULATORY REFORMS PROPOSED BY THE TRUMP ADMINISTRATION RESEMBLE THOSE OF ANY OTHER PRESIDENT IN THE POSTWAR ERA?

Answer: Yes, reforms of the Reagan administration in its first term.

The Facts: The president of the United States from 1981 to 1989 was Ronald Reagan, one of the most popular presidents of the postwar era. During his 1980 campaign for office, he repeatedly regaled crowds of supporters with stories about the burdensome regulations covering so many aspects of life—environmental protection, worker safety, labor policies, employee rights—and business constituents especially complained that government regulations strangled commerce and stifled economic growth. His cry was to "get government off the backs" of the American people, a campaign theme leading to a strong effort at regulatory reform, especially in the early 1980s.

Donald Trump used similar examples of regulatory excess 36 years later in his own presidential campaign. He campaigned on tax cuts, trade reform, promises to crack down on illegal immigration, and "trashing Obama-era regulations" (Clark & Richards, 2020). Shortly after his inauguration he announced that his administration would conduct a "regulatory rollback" of rules and regulations that he and his supporters in the energy industry and other business sectors characterized as overbearing and unnecessary.

Meaning of "deregulation" in the Trump era. As candidate and president, Donald Trump effusively emphasized deregulation, saying: "70 percent of regulations can go." But whereas the Reagan administration's approach to deregulation was orderly and methodical, the approach in the Trump administration was fitful and disorganized.

Rule changes, implementation agencies, and enforcement. The first stage of regulatory reform is action: Congress may eliminate the rule (done infrequently), the president may eliminate it via an executive order (EO), or the agency may suspend it.

The second stage is writing the replacement rule. This step is typically undertaken by the agency or Independent Regulatory Commission (IRC) and involves the Office of Management & Budget (OMB) and its "rule shop," the Office of Information & Regulatory Affairs (OIRA). Previously when agencies suspended application of rules without writing replacements, opposing parties filed suit. The EPA and other agencies have thus emphasized designing rules that can withstand legal scrutiny.

For example, Trump issued a 2017 executive order on "energy independence" that was specifically crafted to provide a legal platform to undo rules such as BLM's methane standards and EPA's Clean Power Plan. Agency staff subsequently claimed they "greatly exceeded" initial targets for slashing regulatory costs, cutting four significant rules for every new one put in place, resulting in $23 billion of net regulatory costs savings in fiscal 2018 (Heikkinen, 2018).

Enforcement of existing rules and regulations also declined markedly during the Trump administration, according to a government watchdog organization called the Environmental Data & Governance Initiative (EDGI). Its first report lambasted the Trump-era EPA for a sharp decline in environmental enforcement (Soraghan, 2018). EDGI updated its 2018 "Sheep in the Closet Report" in 2019 and noted that the number of civil cases was at the lowest level since 1994; criminal enforcement was weak; civil fines collected from polluters were the lowest since 1994; and prosecutions of companies for pollutants and hazardous waste in 2017 and 2018 were the lowest on record. Inspections of infractions and cleanup under Superfund legislation were deficient as well (Fredrickson, 2019). Democratic Senator Maria Cantwell of Washington compared the Trump administration's regulatory agencies to a "cop on the beat" not paying attention to potential lawbreakers (Dillon & Kuckro, 2019).

Cost-benefit analyses and noneconomic effects. Cost-benefit analysis (CBA) is a method used by economists to decide whether a proposed change will promote economic efficiency. The tool has been used in evaluation of environmental policies since the 1980s. In 2018, Resources for the Future (RFF), a Washington, DC think tank, conducted a CBA of six cases in which the Trump administration rolled back regulations on oil and gas production. The report estimated that repealing all rules would save the oil and gas industry $9 billion in 10 years, but cause them to lose $11 billion in other benefits (Krupnick et al., 2018).

The costs of most forms of environmental regulation are narrowly concentrated (businesses pay the bill), while the benefits are widely dispersed: everyone has somewhat cleaner water and air. Moreover, costs are easily numerated in dollars, whereas benefits (e.g., health and safety) often are more difficult to quantify. Stuart Shapiro, a former OIRA staffer, tweeted that the Obama administration "produced net benefits that dwarf those of the Trump administration." Yet some of the Trump administration regulations did produce net benefits, for example, a rule strengthening offshore well control (Lee, 2018).

Can U.S. courts curb the regulatory state? An important question related to regulatory rollbacks is whether agencies or courts have the final word. President Trump's appointment of conservative judges won him great favor with his base, and was a major part of his 2020 reelection campaign. In late 2019, a panel of judges of the 9th U.S. Circuit Court of Appeals dismissed a lawsuit filed by the Center for Biological Diversity (CBD) against a statute erasing many Obama-era environmental regulations. The panel found it lacked jurisdiction to examine the CBD complaint, deferring to the agency (King, 2019). The U.S. Supreme Court has yet to make a final ruling on the degree of deference federal agencies are to be given by the courts. It is expected, however, that the strong conservative/Republican majority on the court will reduce deference to agencies in Democratic administrations (Farah & Richards, 2022).

Changes in public access to rule-making. The Administrative Procedure Act (APA) of 1946 governs the process by which federal agencies develop and issue regulations, and it gives the public opportunities to comment on proposed rule-making. The Trump administration was more vigorous than any previous administration in limiting public access to and commentary on regulatory rule-making. In late 2018, the Interior Department proposed to make it easier to reject Freedom of Information Act (FOIA) requests, including those it deemed "unreasonably burdensome" or requiring "inspection of a vast quantity of material," in response to an "unprecedented surge in FOIA requests and litigation." The streamlining proposal limited the ability of communities and environmental NGOs to overturn or delay air, water, and solid waste permitting, an objective that industry advocates welcomed (Hiar, 2019).

Another environmental NGO, the Center for Western Priorities (CWP), studied 10 major regulation or rule change proposals of Trump's Interior Department, including ones related to OCS policy, endangered species, FOIA requests, and methane leaks. The public submitted millions of comments, more than 95 percent of which were opposed to policy change. However, in 8 of the 10 cases, the department moved forward to make the changes anyway (CWP, 2020).

Biden administration changes: reversing rollbacks. President Biden issued executive orders on the day of his inauguration, starting a comprehensive review process in EPA, Interior, and Energy agencies. A federal appeals court had struck down the Affordable Clean Energy rule, making EPA's work easier. It proceeded to restore vehicle efficiency measures, the role of experts, and of standard setting itself. The Interior Department erased a legal opinion that had softened the Migratory Bird Treaty, restored wildlife protections in clean energy zones, and began resurrection of the department's authority to manage wetlands under the Clean Water Act. An executive order paused all Arctic National Wildlife Refuge oil and gas leasing activities (Richards, 2021). Another first-day announcement from the White House was that ESA rollbacks would be reviewed, which the Sierra Club greeted enthusiastically (Sierra Club, 2021).

The Interior Department evaluated 13 "energy efficiency" rules adopted by the previous administration from a different perspective—whether they would reduce energy use and increase water conservation of appliances such as dishwashers. As in other agencies, evaluation could lead to suspension of the rule, revision, or rescission. The Justice Department restored fines on polluters and renewed the focus on pollution accountability, especially when occurring in communities of color and low income. Also, agency staff identified about 1,400 adverse changes (e.g., removing reference to public comment periods) to federal environmental websites, which were visited and corrected.

Advancing the Biden agenda through law and regulation. The new administration's goals were clean air, land, water; climate change mitigation; and a fast transition to clean (mostly renewable) energy. As mentioned, business groups applauded CBA, and they spoke up in its defense, but rigid use stood in the way of the energy transition. The EPA repealed consideration of expected gains and losses of future Clean Air Act regulations in May 2021. EPA administrator Regan said, "Revoking this unnecessary and misguided rule is proof positive of this administration's commitment to science" (Reilly, 2021), and implicitly to a "whole-of-government" approach to climate action. Early, visible actions going beyond fossil fuels included President Biden's pulling an essential permit for Keystone XL (a campaign promise, notwithstanding loss of 1,000 jobs), and closing loopholes in fracking setbacks.

Biden changes in procedures? In his inaugural address, President Biden said he would take a bold approach to regulation, and he signed a memo called "Modernizing Regulatory Review." OMB was instructed to ensure that "regulatory initiatives appropriately benefit . . . disadvantaged, vulnerable, or marginalized communities" (Brugger, 2021a). Progressives exhorted the president to freeze last-minute Trump administration rules, advice not greatly different from that given the Trump administration about rules made late in the Obama presidency. A different effort was the Department of Transportation's decision to eliminate the Trump administration procedure developed in 2019 to slow down the process of crafting federal regulations, sometimes called the "rule on rules." It might have been used to frustrate Biden administration goals to improve airline safety (Brugger, 2021b). Another procedural change was withdrawing a controversial Trump-era EPA memo questioning the authority of the inspector general, who was investigating a political appointee to the agency. This repaired a rupture in the relationship of an agency and the IG (Bogardus, 2021).

The Biden administration released its first regulatory agenda before the end of its first half-year. Called the "unified agenda," the biennial documents revealed priorities of the administration, which differed substantially from the previous administration. To be expected of a modern Democratic list of problems and concerns, little mention of "deregulation" was made, as compared to the raft of proposals by the Trump administration in its first two years. The attempt to raise standards was manifest in most of the significant actions, which could be called "reregulation."

Comparing the Biden agenda to his three predecessors, about 35 percent of all significant actions were "withdrawals" of Trump administration proposed rules, which is smaller than the 43 percent of the Trump withdrawals of Obama administration rules. Both the Biden and Trump administrations had a greater proportion of withdrawals than the George W. Bush and Obama administrations (21 and 23 percent respectively). Simply put, both Biden and Trump administrations had a stronger focus on regulatory policy than Obama and George W. Bush. Moreover, as power shifted from Obama to Trump and then from Trump to Biden, the "regulatory policy pendulum is swinging with greater magnitude" (Perez, 2021).

Three appraisals put the early steps of the Biden administration in perspective. Two former Trump administration OMB executives feared that CBA would not influence regulatory decisions. They saw that "someone in the new administration not only understands the OMB but knows how to use it . . . (and) the business climate in this country will be made dramatically worse. Forget reregulation; this is the beginning of hyperregulation" (Mulvaney & Grogan, 2021). An observer from the Cato Institute, which typically excoriates Democrats for their anti-market views, believed the Biden administration would resemble earlier industrial policy attempts and be drawn where special interests (such as labor unions) pulled it. They suspected Biden might become "stuck in an old-timer's industrial policy" (Lemieux, 2021). Finally, the *New York Times* (which endorsed Biden for president but was criticized when Biden's early gaffes seemed evidence of incompetence) faintly praised him for having on his team "regulators who believe in regulation . . . (and) do not adamantly oppose the mission of the agencies they aspire to lead" (Eisinger, 2021). Clearly, Biden's administration will pay serious attention to making changes through regulation. However, in one term, facing an evenly divided Congress, it would not be able to return the United States to a model regulatory state.

The NEPA case. The National Environmental Protection Act (NEPA) is the grandfather of American environmental law, and since its passage by the Congress in 1969 (signed by President Nixon in 1970), it has not been revised substantially. The Supreme Court has considered just a few NEPA cases and defended the government from attacks on NEPA (usually by industry). Lower federal courts have required agencies to conduct more detailed environmental impact studies, and this has added to the time required for review (and often extended at industry's request)—now averaging 4.5 years. Prompting revision was President Trump's energy dominance agenda and quest for expedited EIS approvals of projects such as pipelines and OCS drilling (King & Northey, 2018).

In January 2020, President Trump unveiled a "modernization" of NEPA. No longer would direct, indirect, and cumulative effects be discussed separately, and the causal chain linking proposed activities subject to federal agency control (such as OCS drilling) would be shortened. This change effectively removed climate change from consideration in NEPA cases. Second, matters included for NEPA review would need to be "major federal action(s)" whose effects were significant. Time limits of one year for an environmental assessment (EA) and two years for an environmental impact statement (EIS) would be set, as would page limits. Said the president about the revision: "We're going to have very strong regulation, but it's going to go very quickly" (HoganLovells, 2020).

Oil and gas firms, coal, nuclear energy, and the mining industry: all were enthusiastic supporters of the revision. Several Republican activists formed Building a Better America, a coalition to support the NEPA revision. Leading the coalition was Phil Cox, formerly with the Republican Governors Association and Americans for Prosperity. He told gatherings that the greens were "incredibly well funded," and would lobby Congress extensively in opposition to changes in NEPA. His plan was to educate workers, unions, and small business owners and to orchestrate sending of favorable comments to the White House Council of Environmental Quality (CEQ) (Brugger, 2020a).

At hearings of the CEQ, supporters urged expediting EIS reviews to increase employment in energy jobs. Building and agriculture trade groups, chambers of commerce, and trade associations spoke in favor. Environmental NGOs attended in droves and spoke against revisions, arguing that they benefitted polluting industries and endangered public health and safety; fossil fuel was emphasized at the expense of renewables; and climate change was no longer on the table when evidence of its effects was indisputable. Some also said the revisions had been rigged by the Trump administration. Others remarked that NEPA did not hobble projects but led to expeditious reviews, especially in forestry (Heller, 2020). Twenty-two Western governors asked the White House to give them a larger role in reshaping NEPA. Some 170,000 public comments addressed NEPA revisions, an overwhelming majority of which were negative (Brugger, 2020b). Nevertheless, the CEQ completed the revision—one of the boldest deregulatory actions of the Trump administration—by August 2020. Immediately, more than 20 states sued CEQ over the rule, as did a number of environmental NGOs, including Wild Virginia and Cape Fear River Watch, among others.

The incoming Biden administration confronted a major revision of NEPA done by the preceding administration, which had negative impacts on environmental justice and climate change—two high priorities of the new president. Nevertheless, it was the law of the land and would soon be implemented by dozens of agencies and offices. Attorneys for the Department of Justice asked the U.S. district court to remand the Trump overhaul rule to the CEQ because of its flaws, to decide if a new rulemaking process (or amendment) was needed (Joselow, 2021). This put it at odds with environmental NGOs, which wanted the revision repealed instantly—a stance progressive Democrats have taken in many party disagreements. The district court judge declined to freeze the NEPA litigation, sustaining the request of environmental groups.

One month before the deadline for completed implementation of the 2020 NEPA rule, the CEQ began a multistep process. It sent to the White House OIRA a proposal to fix "some of the most critical problems with the previous administration's environmental review process." But in another step, it extended for two years (to 2023) the deadline for agencies to comply with procedural requirements mandated by the Trump NEPA rule (now being implemented by the Biden administration). This produced uncertainty for industry, environmental NGOs, and government agencies. The Trump NEPA revisions were the rules in place. The Biden administration contested the rules, and it had started a new rulemaking process that would take some time to complete (Brugger, 2021c).

The CEQ made final the first phase of changes to rules governing NEPA. Its definition of environmental consequences included "indirect" or "cumulative" impacts, which requires agencies to evaluate how major projects accelerate climate warming and harm communities already affected by pollution. Said CEQ chair Brenda Mallory: "Restoring . . . community safeguards will provide regulatory certainty. . . . Patching these holes in the environmental review process will help projects get built faster . . . and provide greater benefits to people who live nearby." However, the API said changes might jeopardize national security, and other opponents complained about likely increases in red tape (Brugger, 2022).

The regulatory reforms of the Trump administration were numerous and much more penetrating than those of the Reagan administration, and they prompted hostile reactions in several venues—courtrooms, legislative bodies, and on the streets.

FURTHER READING

Bogardus, Kevin. "EPA tossed Trump-era memo that cuffed watchdog," *E&E News*, September 22, 2021.

Brugger, Kelsey. "Coalition forms to defend Trump NEPA overhaul," *E&E News*, January 14, 2020a.

Brugger, Kelsey. "Western governors demand say in NEPA overhaul," *E&E News*, March 25, 2020b.

Brugger, Kelsey. "Biden signals 'new direction' on regulations," *E&E News*, January 21, 2021a.

Brugger, Kelsey. "DOT axes Trump's 'rule on rules'," *E&E News*, March 25, 2021b.

Brugger, Kelsey. "Uncertainty dominates new NEPA approach," *E&E News*, August 18, 2021c.

Brugger, Kelsey. "Biden restores climate to NEPA, undoing Trump's efforts," *Greenwire*, April 19, 2022.

Center for Western Priorities (CWP). "New analysis: Cutting the public out of public lands," January 14, 2020. https://westernpriorities.org/2020/01/new-analysis-cutting-the-public-out-of-public-lands/

Clark, Lesley and Heather Richards. "Trump's 2nd term energy plan? 'Continue what we're doing,'" *E&E News*, August 28, 2020.

Dillon, Jeremy and Rod Kuckro. "FERC eliminates oversight division," *E&E News*, September 16, 2019.

Eisinger, Jesse. "How afraid should corporate America be of Joe Biden?" *New York Times*, February 2, 2021.

Farah, Niina and Heather Richards. "Feds predict NEPA delays after court nixes climate metric," *Energywire*, February 22, 2022.

Fredrickson, Leif. "Update of sheep in the closet report—EPA enforcement record in the Trump administration through fiscal year 2018," Environmental Data & Governance Initiative (EDGI), 2019. https://envirodatagov.org/update-of-sheep-in-the-closet-report-epa-enforcement-record-in-the-trump-administration-through-fiscal-year-2018/

Heller, Marc. "Study counters claim that NEPA hobbles projects," *E&E News*, May 5, 2020.

Hiar, Corbin. "Agency launches major changes for permit appeal process," *E&E News*, November 7, 2019.

HoganLovells. "CEQ's NEPA regulatory overhaul: highlights and predictions," January 15, 2020. https://www.hoganlovells.com/~/media/hogan-lovells/pdf/2020-pdfs/2020_01_15_ceqs_nepa_regulatory_overhaul.pdf

Joselow, Maxine. "Biden asks court for redo on Trump NEPA rule," *E&E News*, March 18, 2021.

King, Pamela. "9th Circuit scraps Congressional Review Act challenge," *E&E News*, December 30, 2019.

King, Pamela and Hannah Northey. "Trump's efforts to spur projects hit NEPA wall," *E&E News*, April 8, 2018.

Krupnick, Alan, Arthur Fraas, Justine Huettemann, and Isabel Echarte. *The Economics of Regulatory Repeal and Six Case Studies*. Washington, DC:

Resources for the Future, November 2018. https://www.rff.org/publications
/reports/the-economics-of-regulatory-repeal-and-six-case-studies/

Lee, Mike. "Benefits of rules outstrip savings in Trump rollback," *E&E News*, November 8, 2018.

Lemieux, Pierre. "Joe Biden's economic agenda: An early appraisal," Cato Institute, Spring 2021. https://www.cato.org/regulation/spring-2021/joe -bidens-economic-agenda-early-appraisal

Mulvaney, Mick and Joe Grogan. "Biden gives regulators a free and heavy hand," *Wall Street Journal*, January 26, 2021.

Perez, Daniel. "The Biden administration's first unified agenda," Regulatory Studies Center, June 16, 2021. https://regulatorystudies.columbian .gwu.edu/biden-administration's-first-unified-agenda

Reilly, Sean. "EPA revokes Trump-era barrier to climate rules," *E&E News*, May 13, 2021.

Richards, Heather. "Executive order will pause all ANWR oil activities," *E&E News*, January 20, 2021.

Sierra Club. "On day one, President Biden announces review of Endangered Species Act rollbacks," January 20, 2021. https://www.sierraclub .org/press-releases/2021/01/day-one-president-biden-announces-review -endangered-species-act-rollbacks

Soraghan, Mike. "Trump team reluctant to pursue polluters—report," *E&E News*, November 10, 2018.

"Trump updates rule-busting agenda," *E&E News*, October 17, 2018.

Q29. HAVE FEDERAL DEPARTMENTS AND REGULATORY AGENCIES BEEN "CAPTURED" BY THE INDUSTRIES AND BUSINESS INTERESTS THEY ARE RESPONSIBLE FOR REGULATING?

Answer: Yes, in some cases.

The Facts: The "capture thesis" has a history of more than 100 years in the United States. The issue concerns business–government relations in capitalist society, which change over time. It is the proverbial Goldilocks question: Is the relationship between government and business too hot (close)? Too cool (distant)? Or just right? If relationships between business (especially leaders of dominant sectors in the economy) and government agencies are too close, citizens worry that decisions are not made in the public interest and may be corrupt. In the last generation, remedies have been devised to balance the government–business relationship (McBeath, 2016; Rex, 2018).

This question explores the extent to which the capture thesis applies to the Department of Energy (DOE), the Department of the Interior (DOI), the Environmental Protection Administration (EPA), and the Pipeline and Hazardous Materials Safety Administration (PHMSA). Space limitations allow only review of the industry linkages of chief personnel and a few examples of apparent capture in the Trump administration (keeping in mind that before becoming president, Trump had been a champion of American corporations). The National Association of Manufacturing (NAM) applauded the Trump administration for acting on 85 percent of its rulemaking requests and responding to 64 percent of wish list items. As one NAM lobbyist said, "There's no denying that we've been effective advocates for our members and have had a great partner in the Trump Administration, but what too often gets lost in the conversation is that we aren't simply saying 'no' to regulation . . . what we are pushing for is smarter, better regulation" (Heikkinen, 2019). But Amit Narang, regulatory policy advocate for the environmental NGO Public Citizen, offered a more critical reaction to that record: "I think it's a very clear indication of how captured the Trump administration is by corporations."

Department of Energy. Trump's first energy secretary was Rick Perry, former two-term governor of Texas, America's largest oil/gas state. Perry was selected for the position in part because of his reputation as a forceful advocate for fossil fuels. His assistant, Doug Matheney, had worked for Americans for Prosperity, the Koch Brothers, and as Ohio director of Count on Coal (a PR campaign funded by the coal industry). Carl Icahn, a senior administration advisor, oil refinery investor, and billionaire, influenced regulatory policy pertaining to the energy industry as well (Gaby, 2017).

Writing in *The Guardian* a year before the 2020 election, journalist Peter Stone listed leaders of the fossil fuel sector who had contributed to the Trump Victory Committee—Rob Murray (Murray Energy, coal), Jeff Miller (fossil fuel lobbyist), Kelcy Warren (head of Dakota Access pipeline construction), Harold Hamm (CEO, Continental Resources), and Joe Craft (CEO, Alliance Resource Partners). He quoted Jerry Taylor of the nonpartisan Niskanen Center: "It's hard to identify any industrial sector that has ever had this much success with *any* administration in modern history. The fossil fuels industry has gotten nearly every single last item on its wish list from the Trump administration" (Stone, 2019).

Department of the Interior. The Interior Department manages greater than 500 million acres of land, 1.7 billion acres of ocean floor, the animals and plants living there, and all the liquid and hard rock minerals lying below the surface. Two of its top leaders in 2020—David Bernhardt, secretary, and William Pendley, acting head of the Bureau of Land Management (BLM)—had been charged with conflicts of interest. Bernhardt had held posts in high-profile fossil fuel firms, such as Halliburton Energy, the Independent Petroleum Association of America, and the U.S. Oil & Gas Association. Pendley had a long association with nonrenewable energy interests and had long been a vocal advocate of selling off public lands to private business interests. He compared government regulation to tyranny, and once compared climate change to a "unicorn" because "neither exists" (Friedman & O'Neill, 2020).

EPA. President Trump's first appointment to head EPA was Scott Pruitt, who held unrecorded meetings with the Thomas Hill Energy Center, a company in noncompliance with the Clean Water Act since the 1990s (Hiar, 2019). Pruitt and his replacement at EPA, Andrew Wheeler, also helped a Virgin Islands refinery acquire environmental permits allowing it to reopen despite a history of oil spills, contaminated groundwater, and prospective endangerment of federally protected coral). The Center for Biological Diversity (CBD) called the refinery an "environmental monster that Trump officials brought back to life" through the EPA's "well-documented customer service approach to fossil fuel companies" (Brugger, 2021). The CBD's petition to deny the refinery permit was lodged in the EPA on environmental justice grounds after Biden took office because pollution from the refinery afflicted a predominantly Black, Latino, and low-income area. In June 2021 the refinery closed its operations indefinitely.

PHMSA. This agency was created by Congress in 2005 to improve pipeline safety; it is housed in the transportation department. Since its creation, the agency has been accused of lax oversight of pipeline companies and of being captured by the industry it is supposed to monitor. An inspector general's (IG) report remarked that "industry and PHMSA are not sufficiently separate," and its survey of employees' perceptions indicated that agency leaders were reluctant to seek penalties against companies for infractions (although in its defense, the agency is limited in the range of fines it is authorized to impose). Still, the IG report found the agency maintained a safety-oriented "culture" (Soraghan, 2021).

These ties between agency officials and the industries they are supposed to police appear suspiciously close in some instances. However, "capture" is a potential problem for all U.S. administrations. When President Eisenhower appointed Charles E. Wilson to serve as secretary of defense, Senator Hendrickson asked him whether his shares in the corporation he headed (General Motors) were more important than his loyalty to the United States, he said: "What is good for General Motors is good for the country, and what is good for the country is good for General Motors (GM)" (Terrell, 2016). Replacing GM with "fossil fuels" in the case of the Trump administration, and "renewables" in the case of the Biden administration brings this relationship up to date.

Biden administration. Some of the major critics of Trump administration energy and environmental policy were appointed to high-level positions in the Biden administration. President Biden's long experience in the Senate and as vice president for eight years assisted in the selection and vetting of nominees, and in the expeditious placement of nonconfirmed appointees. Timing was slowed by the second impeachment of President Trump, the pandemic, and a continued high level of partisan polarization. President Biden's nominees for leadership of Energy, Interior, and EPA were confirmed and with some Republican votes, but several faced intense scrutinies (Bogardus, 2021). For example, suspicions clustered around former Michigan governor Jennifer Granholm's financial investments (especially in electric vehicles, lithium mining interests, and renewables). In some cases, the policy and ideological stances of nominees were the apparent target of critiques, but the underlying focus was their distance from clean energy/renewables (or fossil fuels in the Trump presidency).

FURTHER READING

Bogardus, Kevin. "Biden's EPA noms face sluggish confirmation," *E&E News*, August 19, 2021.

Brugger, Kelsey. "Greens challenge permit for troubled Virgin Islands refinery," *E&E News*, February 3, 2021.

Friedman, Lisa and Claire O'Neill. "Who controls Trump's environmental policy?" *New York Times*, January 14, 2020.

Gaby, Keith. "Conflicts of interest taint Trump's environmental picks: Top 7 examples," Environmental Defense Fund, May 10, 2017. https://www.edf.org/blog/2017/05/10/conflicts-interest-taint-trumps-environmental-picks-top-7-examples

Heikkinen, Niina. "Trump fulfills most of trade group's wish list," *E&E News*, May 10, 2019.

Hiar, Corbin. "Inside top EPA officials' secret polluter meetings," *E&E News*, April 15, 2019.

McBeath, Jerry A. *Big Oil in the United States: Industry Influence on Institutions, Policy, and Politics.* Denver: Praeger, 2016.

Rex, Justin. "Anatomy of agency capture: An organizational typology for diagnosing and remedying capture," *Regulation & Governance*, July 31, 2018.

Soraghan, Mike. "PHMSA staff concerned about agency-industry ties—IG," *E&E News*, January 19, 2021.

Stone, Peter. "'Swampy symbiosis': Fossil fuel industry has more clout than ever under Trump," *The Guardian*, September 27, 2019.

Terrell, Ellen. "When a quote is not (exactly) a quote: General Motors," Library of Congress (Blog): Inside Adams, Science, Technology & Business, April 22, 2016.

Q30. HAVE FEDERAL–STATE CONFLICTS OVER ENERGY AND ENVIRONMENTAL POLICIES INCREASED IN THE LAST DECADE?

Answer: Yes.

The Facts: Article VI, Clause 2 of the U.S. Constitution avows that the Constitution, laws of the United States, and all treaties made under national authority "shall be the supreme law of the land . . . and the judges in every state shall be bound thereby." The next clause continues: "Senators and Representatives . . . and the members of the several State Legislatures, and all executive and judicial Officers, both of the United States and of the several States, shall be bound by Oath or Affirmation to support this Constitution." It was not until after the Civil War that this language was used to enforce the supremacy of federal laws and treaties over state laws.

The commerce clause refers to Article 1, Section 8, Clause 3 of the Constitution, which gives Congress the power "to regulate commerce with foreign nations, and among the several states, and with the Indian tribes." Traditionally, the commerce clause has been seen as granting Congress authority to regulate commerce with foreign nations and among the states, while simultaneously restricting the regulatory authority of the states to discriminate against residents of other states, but the phrasing pertaining to the states in the clause is often called "dormant." Case Western Reserve law professor Jonathan Adler wrote that with respect to overtly discriminatory or protectionist measures, the dormant commerce clause was alive and well: "That certainly means that states that are adopting arguably discriminatory climate policies will have to pay attention to this concern" (King, 2019b).

During the 1970s, a political movement known as the Sagebrush Rebellion swept across a number of Western states containing large amounts of federal public land. Ranchers and other members of the rebellion bitterly complained that federal land use policies were unfair and unresponsive to their needs. The Sagebrush Rebellion itself eventually faded away as a political movement, but anger and dissatisfaction with the land use policies of the Bureau of Land Management (BLM) and other federal agencies charged with managing public lands remained high in many parts of the West.

Since then, several factors have revived state opposition to federal supremacy: increased emphasis in the last several administrations on energy independence; the very rapid growth of renewables and subsequent conflict between those industries and fossil fuels; and continued partisan conflict, especially noticeable during federal and state power shifts from one party to the other.

Types of federalism. Since the Depression-fighting New Deal transformed American society in the 1930s, the federal government has exerted tremendous influence over national social and economic policy, including energy and environmental law and policy. It has been able to do so in part by relying on *fiscal federalism*—the superior revenues of the federal government. Still, three other versions of federalism have also come into play: 1) *Dual federalism*, the situation in which both federal and state governments are sovereign, remains in effect since the framing of the Constitution in 1789. The states have areas or fields of autonomy—for example, legalization of marijuana while it remains illegal under federal law. States singly or collectively may attempt to use their autonomy to weaken or disobey federal policies they disagree with. 2) *Cooperative federalism* expresses the belief that all levels of government should work together to resolve common problems, such as economic growth or environmental protection. The federal government is a partner that negotiates with states (also tribes, local governments) about the development of law and policy and its implementation. 3) *Coercive federalism* refers to times when the federal government uses power unilaterally, attaching many regulations to federal grants, issuing mandates on states and preempting state policies. For instance, the federal government orders the states to protect the environment and mitigate negative externalities such as pollution (Kincaid, 2017). The issue this question addresses is whether federal regulations are a *floor* (meaning states may exceed the action requested, such as require cars to further reduce emissions) or a *ceiling* (meaning states may not require more than the federal standard).

The tensions between state and federal authority can be seen in examining energy implications pertaining to: 1) standards of the Clean Water Act (CWA) and Waters of the United States (WOTUS); 2) standards of the Clean Air Act (CAA); 3) Western states' attempts to acquire Bureau of Land Management (BLM) lands; and 4) Outer Continental Shelf (OCS) and coastal energy development.

Clean Water Act (CWA) and WOTUS permitting issues. Administering most of the U.S. pollution control and water quality programs is the EPA, which sets wastewater standards for industry and controls discharges of pollutants from sources such as pipes or ditches into navigable waters (EPA, 2020). Under the CWA, Congress extended EPA's control to navigable waters, and toward the end of the Obama administration, EPA developed a rule defining the Waters of the United States broadly. The Trump administration EPA recodified this into a narrower Navigable Waters Protection Rule (NWPR) (EPA, 2019). Both CWA and WOTUS allowed states to weigh in on federal permits; however, the Trump administration proposed to dramatically curtail the ability of states to stop new construction and other development for environmental reasons.

The Trump administration took this stance after energy companies complained that state governments in New York and Washington blocked permits to allow pipeline construction and other new projects. The Western Governors' Association expressed support for the position taken by New York and Washington. It asked Trump "to reject any changes to agency rules, guidance or policy that may diminish, impair, or subordinate states' well-established sovereign and statutory authorities to protect water quality within their boundaries" (Wittenberg, 2019a). Instead, Trump told EPA to override states' objections and approve pipeline construction, a decision praised by the energy industry and condemned by environmental groups.

California and Oregon then invoked Section 401 of the CWA to review a proposed hydroelectric project opposed by environmental groups including California Trout and Trout Unlimited (Gilmer, 2019). Section 401 of CWA gives states the authority to allow, deny, or waive certification of proposed federal leases/permits that may discharge pollutants into U.S. waters. Section 401 thus stands as an important check on federal regulatory power.

Trump responded by signing two executive orders to increase federal control over pipeline approvals while limiting states' ability to block projects. The executive orders sought expanded energy production, and amended safety rules for natural gas export facilities. The Western Governors' Association criticized the orders, calling the Trump administration's position "fair-weather federalism" that flouted traditional Republican support for states' rights (Wittenberg & Brugger, 2019). Finally, in an end-of-administration action, Trump's EPA approved a request from the Texas Commission on Environmental Quality to issue permits to oil and gas firms to dispose of treated wastewater into rivers and streams—another decision denounced by environmental groups (Northey & Lee, 2021).

The WOTUS permitting question was whether states could provide a higher degree of environmental protection for its natural water resources than federal standards allowed. In 2008, the California water board sought to establish standardized rules defining which wetlands were protected by the state and what was required of factories, farmers, ranchers, housing developers, and others to obtain operating permits for. Updating rules took a decade because of the need to establish baselines of wetlands quality (essential for mitigation purposes) and to ensure that regulators understood the impacts. State water resources control board officer Jonathan Bishop said, "We need to have clear rules for Waters of the State so that no matter what happens with WOTUS, we are still protecting these important features in California" (Wittenberg, 2019b). When the state rules proved tougher than the federal standard, Trump administration EPA head Andrew Wheeler threatened to withdraw California's federal highway funds for noncompliance. Other states, such as Colorado and Arizona, also were concerned that the Trump administration standard would be difficult to change, and crafted legislation to regulate and protect their wetlands and streams (Northey, 2021).

Clean Air Act (CAA) and related matters. The CAA regulates air emissions from stationary and mobile sources; it authorizes EPA to establish standards to protect public health and welfare and to regulate emissions of hazardous air pollutants. The act directs states to develop state implementation plans (SIPs), working with state environmental quality agencies (EPA, 2019). The CAA became a political flashpoint after the 2018 midterm elections, when some newly elected Democratic governors (e.g., Illinois, Connecticut) proposed climate change and clean energy actions opposed by the Trump administration (Tomich, 2018). (After the 2020 election, 12 governors asked President Biden to set up a regulatory framework that would create good market conditions for zero-emissions vehicles [Skibelt, 2021].) A second air quality issue concerned California's cap-and-trade agreement with the Canadian province of Quebec to limit greenhouse gas releases and set an emissions market. The Trump administration called the agreement an "independent foreign policy" in violation of federal law.

Meanwhile, Wheeler issued accusations that the California Air Resources Board (the state's main air quality agency) had failed to submit dozens of SIPs and failed to rectify non-attainment area pollution. The Clean Air Act is applied to sections/divisions/areas of U.S. states, and its objective is to ensure that all of the divisions meet National Ambient Air Quality (NAAQS) standards. Those that fail to do so are called nonattainment areas, with respect to six criteria pollutants (e,g., sulfur dioxide, ozone), and regulators attempt to bring them up to the standard.

Wheeler also claimed that California had "the worst air quality in the United States." CARB director Mary Nichols challenged the "many inaccuracies and misleading statements" from Wheeler about the state's air quality, reminded him that "CARB was established years before U.S. EPA came into existence," and claimed that while California had done its part to combat air pollution, "EPA has not done its part" (Joselow, 2019). In 2021, the Biden administration moved quickly to restore this regulatory authority to California, a move in keeping with its wider climate change strategy (Reilly, 2021).

Western states' attempts to acquire BLM lands. The Bureau of Land Management (BLM) manages more surface land (245 million acres, approximately 10 percent of America's land base) and subsurface mineral estate (700 million acres) than any other U.S. agency. Its mission is to administer public lands "on the basis of multiple use and sustained yield" of resources. Uses include conventional energy development (oil and gas, coal), renewables (solar, wind, etc.), hard rock mining (gold, silver, etc.), timber harvesting, and outdoor recreation. It also has conservation objectives such as preserving designated landscapes (BLM, 2020).

During his 1980 presidential campaign, Ronald Reagan declared himself a supporter of the Sagebrush Rebellion, a grassroots movement that wanted federally owned lands in the West to be sold or parceled out to states and private developers. Nearly 40 years later, during the 2016 election campaign, Donald Trump opposed federal land sales to states, but his position on the issue in the presidential race of 2020 was more ambiguous. The Reagan campaign of 1980 and the Trump campaign of 2020 were similar in their enthusiasm for doing away with environmental regulations that limited energy development and other industrial activity. Both favored delegating Endangered Species Act (ESA) authority to the states, streamlining NEPA reviews for energy projects, opening up public lands containing "strategic minerals," and increasing access of off-road vehicles to wilderness areas on public lands. These goals were supported by Western states controlled by pro-development Republican legislatures and governors, but they met with strong opposition from environmental groups and Democrat-led states.

OCS and coastal development and planning. The American Outer Continental Shelf (OCS) is rich in oil and gas resources and a fertile environment for the development of renewables such as wind and solar. The National Oceanic & Atmospheric Administration (NOAA) manages coastal resources through a cooperative agreement with the Bureau of Ocean Energy Management (BOEM). It is the Interior Department agency in charge of the OCS oil and gas leasing program. States develop coastal management programs to administer and balance competing uses of the coastal zone, and under this state program, consistency determinations are arrived at with respect to proposed energy resource developments (CZMA, NOAA, 2020). In 1972, Congress passed the Coastal Zone Management

Act (CZMA) to give individual states more input on coastal protection measures. The CZMA gave coastal states responsibility to protect and promote coastal resources, and it included provisions that states could use to flag federally approved projects "inconsistent" with state plans.

The Trump administration planned an aggressive and expansive OCS energy development effort, and sought to limit state authority over development along their shores. Democratic governors of nine states urged NOAA not to streamline federal determinations on whether offshore energy projects aligned with state interests. The director of the State Energy & Environmental Impact Center, which supported multistate legal challenges to the Trump administration's aims, claimed that "the administration is looking to preemptively disempower states from exercising their rights to object to non-complying federal projects under the Coastal Zone Management Act" (King, 2019a). Many of the same states also contested nearly every attempt by EPA and other federal environmental agencies to eliminate rules such as the Clean Power Plan and Clean Water Rule. As of mid-2020, they had won 80 percent of the cases they challenged (King & Jacobs, 2020).

The states in support of Trump administration policy rollbacks were also most likely to oppose Biden administration proposals. A survey of recent actions by the American Council for an Energy-Efficient Economy (ACEEE) identified a large political split. States that had supported Biden in the 2020 presidential campaign led the way in endorsing energy efficiency and clean energy policies (extending to policies on motor vehicles, buildings and appliances). States that supported Trump in the 2020 election, on the other hand, were reluctant to endorse clean vehicles standards, building efficiency and environmental justice, but strongly supported fossil fuel industries (Behr, 2022).

FURTHER READING

Behr, Peter. "Report shows red-blue state divide on clean energy," *Energywire*, February 3, 2022.

E&E Staff. "5 state battles to watch," May 16, 2019.

Gilmer, Ellen. "Groups push Supreme Court to take up state permitting fight," *E&E News*, August 27, 2019.

Joselow, Maxine. "Calif. air regulator fires back at Wheeler," *E&E News*, October 10, 2019.

Kincaid, John. "Introduction: The Trump interlude and the states of American federalism," *State and Local Government Review*, Vol. 49, no. 3 (September 2017): 156–69.

King, Pamela. "State AGs critique coastal management changes," *E&E News*, April 29, 2019a.

King, Pamela. "Commerce clause ruling packs punch for state energy policy," *E&E News*, June 27, 2019b.

King, Pamela and Jeremy Jacobs. "States lead court fight against Trump. They're winning," *E&E News*, May 26, 2020.

Northey, Hannah. "States face quagmire in wake of Trump rule," *E&E News*, April 26, 2021.

Northey, Hannah and Mike Lee. "Critics fume after EPA hands oil permitting to states," *E&E News*, January 19, 2021.

Reilly, Sean. "EPA to restore Calif. authority over tailpipe emissions," *E&E News*, April 26, 2021.

Skibelt, Arianna. "Governors to Biden: Stop sales of gas-powered cars by 2035," *E&E News*, April 27, 2021.

Tomich, Jeffrey. "Blue wave in Midwest raises hopes for more green energy," *E&E News*, November 8, 2018.

U.S. Department of Commerce, NOAA. "CZMA: Federal consistency overview," 2020. https://coast.noaa.gov/data/czm/consistency/media/federal-consistency-overview.pdf

U.S. Department of the Interior, Bureau of Land Management (BLM). "What we manage," March 2020. https://www.blm.gov/about/what-we-manage/national

U.S. Environmental Protection Agency. "WOTUS step one–Repeal," 2019. https://www.epa.gov/wotus/wotus-step-one-repeal

U.S. Environmental Protection Agency. "Summary of the Clean Water Act," 2020. https://www.epa.gov/laws-regulations/summary-clean-water-act

U.S. Environmental Protection Agency. "Clean Air Act Text," 2021. https://www.epa.gov/clean-air-act-overview/clean-air-act-text

Wittenberg, Ariel. "Governors to Trump: Don't meddle in state permit reviews," *E&E News*, February 1, 2019a.

Wittenberg, Ariel. "Wheeler to Calif.: Clean up San Francisco sewers," *E&E News*, September 26, 2019b.

Wittenberg, Ariel and Kelsey Brugger. "EPA focused on 4 projects in crafting state permit rule," *E&E News*, August 20, 2019.

6

❖

Trade-Offs

The rapid post-Civil War settlement of the American West and consequent national economic development had lasting and far-reaching effects on land, water resources, ecosystems, and the growing human population. Industrialization under the philosophies of "manifest destiny" and human dominance of nature were reflected through the nineteenth and early twentieth century in government policy: economic development aimed at improving the land by exploiting natural resources for economic growth and building national political and military strength (McBeath & Rosenberg, 2006). Although there were many state- and local-level stirrings against this unsustainable level of resource extraction, it was not until the emergence of the Progressive movement (roughly 1890 to 1920) that meaningful limits began to be placed on America's prodigious appetite for land and the natural resources contained therein. The Progressive movement inspired conservation efforts to conserve valuable natural resources from complete depletion and to protect charismatic landscapes and megafauna from destruction.

Congress set aside parts of the public domain in national parks, forests, grazing lands, wildlife refuges, and recreational areas. During the Great Depression, the federal government promoted conservation using methods and programs (such as the Civilian Conservation Corps) that also provided economic aid to suffering Americans (Merchant, 2005). Conservation leaders also popularized new principles, such as multiple use and sustained yield, that became incorporated into environmental law. Under the

Multiple-Use Sustained-Yield Act (1960), for example, forests and other resources on public lands were to be managed not primarily for economic gain, but for a balanced combination of "outdoor recreation, range, timber, watershed, and wildlife and fish purposes." Sustained yield, meanwhile, was a principle that emphasized that extraction of resources from public lands should be done on a sustainable basis, so that future generations would also be able to benefit from the resource.

Later, in the 1960s, Congress passed a series of public land policies, including the Wilderness Act (1964), the Land and Water Conservation Fund Act (1964), and the Wild and Scenic Rivers Act (1968), in response to growing public concern that America was despoiling and polluting its land, air, and water resources (Kraft, 2015).

Environmental historian Samuel Hayes remarks that the period between the conservation movement that arose early in the twentieth century and the environmental movement of the late 1960s/1970s reflected a shift of target from the efficient use of natural resources to a quality of life expressing "beauty, health, and permanence" (Hays, 1982). The latter years of this evolution coincided with a series of postwar events and crises that also prompted significant cultural change, including the civil rights movement, the anti-Vietnam War mobilization, and the women's movement.

Pollution events were another stimulus to new and more robust government action in the realm of environmental protection, as oil spills, fires on lakes and rivers, and land degradation aroused fear, anger, and shame. Rachel Carson's book *Silent Spring* (1962) energized people about the impact of the chemical DDT and other insecticides on animal populations. These events and movements expanded the agenda of issues, which crystallized in the full-blown American environmental movement (Vig & Kraft, 2010).

Environmental protection regulatory regime. The inaugural work of environmentalism was the National Environmental Policy Act (NEPA), adopted by the Congress in 1969 and signed into law by President Nixon in 1970 (see Q28). NEPA ushered in a raft of environmental laws that were "truly extraordinary for a political system where the norm is incremental policy change" (Kraft, 2015). They included the Clean Air Act (CAA, 1970, amended in 1990), the Clean Water Act (CWA, 1972), the Safe Drinking Water Act (SDWA, 1974), and the Endangered Species Act of 1973 (McBeath, 2004). Toxic waste legislation was enacted as well, along with cleanup of hazardous-waste sites in the Superfund (Comprehensive Environmental Response, Compensation, and Liability Act, or CERCLA). Each act created standards and, through a top-down process (called in the literature "command-and-control"), dominated by the federal government, told business corporations and local governments what they needed to do to pass muster. The implementing agency for most of the laws was the Environmental Protection Agency (EPA), established by an executive order of President Nixon in 1970.

The consensus and bipartisan coalition initiating this decade of environmental law soon fractured as new national challenges—two oil shocks in the 1970s, stagflation, and foreign conflict—preoccupied leaders. Also, the election of Ronald Reagan in 1980 brought a new aura of conflict into both energy and environmental policymaking (see Q28). Yet conflict inspired strong defense of environmental goals in the growing third (nonprofit) sector. Environmental NGOs such as Greenpeace, the Wilderness Society, and the Nature Conservancy were established and grew in membership as environmental goals were identified and pursued. As in previous chapters, the scope of coverage here includes the Obama and Trump terms of office and the first year of the Biden administration. Seven questions focus discussion. Q31 asks whether the Clean Air Act, including amendments passed in 1990 that strongly emphasized enforcement of air pollution laws, has improved U.S. air quality. The next two questions consider atmospheric emissions, and whether they have worsened or not since the early 2010s. Q32 treats the CAFE standards on auto emissions, and Q33 examines trends in public attitudes toward fossil fuel use in the context of climate change.

Q34 evaluates the quality of the water people drink and use in commercial and industrial processes, as well as water quality in natural ecosystems. Q35 analyzes the impact that oil and gas drilling and production (and also hard rock mining) have had on threatened and endangered species and their habitats. Q36 asks whether NGOs such as the Sierra Club, Environmental Defense Fund, and Greenpeace have become more extensively involved in litigation than in their first decades of existence. Q37 asks if the so-called Green New Deal, an idea crafted and supported primarily by liberal Democrats, is a feasible model for environmental change.

FURTHER READING

CAA. "Summary of the Clean Air Act," September 28, 2021. https://www.epa.gov/laws--regulations/summary-clean-air-act

Carson, Rachel. *Silent Spring.* New York: Houghton Mifflin, 1962.

CWA. "Summary of the Clean Water Act," October 21, 2021. https://www.epa.gov/laws-regulations/summary-clean-water-act

Dietz, Thomas and Paul Stern, eds. *New Tools for Environmental Protection: Education, Information, and Voluntary Measures.* Washington, DC: National Academy Press, 2002.

Hays, Samuel. "From conservation to environment: Environmental politics in the United States since World War II," *Environmental Review*, Vol. 6, no. 2 (Fall 1982): 14–41.

Kraft, Michael E. *Environmental Policy and Politics*, 6th ed. New York: Routledge, 2015.

McBeath, Jerry. "Management of the commons for biodiversity: Lessons from the North Pacific," *Marine Policy*, no. 28 (2004): 523–39.

McBeath, Jerry and Jonathan Rosenberg. *Comparative Environmental Politics*. Dordrecht, The Netherlands: Springer, 2006.

Merchant, Carolyn. *The Columbia Guide to American Environmental History*. New York: Columbia University Press, 2005.

Rosenbaum, Walter A. *Environmental Politics and Policy*, 11th ed. Washington, DC: Sage/ Congressional Quarterly Press, 2020.

SDWA. "Summary of the Safe Drinking Water Act," September 28, 2021. https://www.epa.gov/laws-regulations/summary-safe-drinking-water-act

Vig, Norman and Michael Kraft. *Environmental Policy: New Directions for the Twenty-First Century*, 8th ed. Washington, DC: Sage/Congressional Quarterly Press, 2010.

Q31. HAS THE CLEAN AIR ACT AND SUBSEQUENT AMENDMENTS IMPROVED U.S. AIR QUALITY IN THE LAST DECADE?

Answer: In the decade up to 2016, U.S. air quality in regard to emissions of most pollutants improved; evidence is mixed as to whether this condition continued to 2022.

The Facts: Many chemical compounds in the planet's atmosphere act to trap radiation close to the surface, warming the earth, much as a greenhouse warms growing plants. Some of the gases such as water vapor present no problem. Another, ozone, is problematic only when close to the earth; if ozone levels are insufficient in the upper stratosphere, the sun's UV rays can burn skin. The major gases the Energy Information Administration (EIA) includes in American registries of greenhouse gas (GHG) emissions—the leading causes of climate change—are carbon dioxide (CO_2), methane (CH_4), nitrous oxide (N_2O), and industrial toxins such as perfluorocarbons (PFCs) (EIA, 2021).

The definition of the American Geosciences Institute (AGI) concentrates on all combustible fuels that emit harmful gases and particles when burned to provide energy. It distinguishes stationary sources (e.g., oil and gas fields, oil refineries and gas processing plants) from mobile sources (e.g., cars, trucks, trains, airplanes). Mobile sources comprise the larger proportion of pollution in the United States. AGI's list of major air pollutants includes four gases (carbon monoxide, nitrogen oxides, sulfur oxides, ozone), particulate matter (PM), and lead (Pb) (AGI, 2018). The Environmental Protection Agency (EPA) sets air quality standards for these six major air pollutants (called "criteria" air pollutants).

Changes in PM 2.5 (soot) and other "criteria" air pollutants. In late 2019, two Carnegie Mellon economics professors reported that after declining by 24 percent from 2009 to 2016, fine particulates (PM 2.5) increased by 5.5 percent between 2016 and 2018 (Clay & Muller, 2019). (They based their conclusions on actually measured pollution levels, not emissions estimates.) These findings contradicted Trump administration officials who insisted that "the environment is getting cleaner" despite its moves to relax pollution regulations (Bote, 2019). Trump EPA officials objected, stating that particulate levels had declined between 2016 and 2018, but the agency's data over Trump administration years showed a smaller rate of improvement for particulates than gases (EPA, 2019).

When Trump claimed that "We have the cleanest air in the world in the United States and it's gotten better since I'm president," the *New York Times* compared American fine particulate pollution with that of other rich (OECD) countries and found that the United States ranked tenth, behind countries such as Canada, Australia, New Zealand, and several European states. Furthermore, over 110 million Americans lived in cities and counties with unhealthy levels of pollution (Popovich, 2019). Another comprehensive air quality study produced by the Yale/UNEP Environmental Performance Index (EPI) ranked the United States 16th out of 180 countries in overall air quality—better than most countries, but not as good as other developed countries such as Finland, Sweden, Iceland, New Zealand, Canada, France, Portugal, and the United Kingdom. The American Lung Association (the environmental NGO paying greatest attention to air pollution issues) described U.S. air quality conditions as unhealthy for those suffering from asthma, allergies, respiratory, and cardiovascular problems.

The Carnegie Mellon professors mentioned three factors explaining their findings: First, economic activity had increased, resulting in more factories producing goods, more smokestacks emitting pollutants, and more cars and trucks on the roads. Second, wildfires had become more severe, propelling large and small micron particulates into the air. Third, CAA enforcement activity had declined (noted in Q29). Assuredly, a large part of the reduction in air pollution was due to the implementation of command-and-control methods including shutting down smokestacks and fining owners. Yet while the general public benefitted from air quality regulations, industry vociferously objected to the added business costs of meeting these rules. Although both Republican and Democratic administrations employed market-based incentives (MBIs) to lessen use of command-and-control methods, the efforts dragged and enforcement slowed. The Trump-era EPA's rosy findings that air quality had improved from 2016 to 2018 were challenged, leading to an investigation by a panel of the National Academies (Reilly, 2021a).

Clean Power Plan (CPP) of the Obama administration. The purpose of the CPP was to use governmental authority conferred by the Clean Air Act to significantly reduce carbon dioxide emissions driving climate change. Domestically, air pollution could be reduced; globally, the Obama administration's credentials as a leader in the fight against climate change would be enhanced. Power plants were the target, both coal- and gas-fired. The Obama-era EPA drafted the CPP following White House instructions. The rule spelled out GHG reductions for states; failing to reach the threshold would result in EPA's imposing an implementation plan (Hildreth, 2018). Prominent national environmental groups such as the Natural Resources Defense Council (NRDC) and the Sierra Club supported the CPP strongly, calling it the single most important way America could fight climate change.

The initiative, however, ran into fierce opposition from the fossil fuel industry and its allies. First to object to the CPP proposal was Murray Energy, the largest U.S. privately owned coal mining company (whose owner was a strong lobbyist for coal and fossil fuels). Then a dozen coal-producing states sued the EPA. Coal miners demonstrated against potential losses of jobs and higher electricity costs for consumers.

After the rule was issued, 27 states petitioned the court for an emergency stay, claiming that the new regulations would result in closures of numerous coal plants. The key issue they raised to combat the Obama administration was the constitutionality of the CPP, because it was based on an aberration in congressional procedure. (The Senate and House passed different versions of the CAA 1990 amendments, but the Constitution required that they be identical.) In 2016, the conservative-majority Supreme Court stayed implementation of the rule, and Obama's term ended before the constitutional issue was resolved. The Trump administration immediately abandoned the CPP along with all other Obama administration efforts to reduce greenhouse gas emissions. By then the EPA was led by Scott Pruitt, who had been one of the leading critics of CPP as attorney general of Oklahoma (Hildreth, 2017).

Although the CPP fell short of full implementation, several objectives were met. Hundreds of coal plants closed, natural gas prices dropped, and the energy source statistics in 2019 showed a reversal of the roles of coal and natural gas in the energy mix. Prices of wind and solar power declined, and power plant operators warmed to using renewables. Utilities had surplus electricity as energy efficiency measures took effect (Kuckro, 2019).

Affordable Clean Energy rule of the Trump administration. The Trump administration had a deregulatory mindset when it came to the fossil fuel industry, but EPA was required to regulate power plants because of the "endangerment ruling" (a legal opinion that GHG emissions were a threat to public health and had to be reduced). Court challenges to CPP indicated that EPA's best option was "inside-the-fence-line" measures, such as replacing or upgrading boilers at the plants. Unlike the CPP, the Affordable Clean Energy (ACE) rule adopted by the Trump administration in August 2018 opened the gateway for a resurgence of coal if prices of natural gas or renewables rose (Friedman & Plumer, 2017).

The trajectory of the ACE rule resembled that of CPP. Environmental NGOs protested at public hearings, critiquing ACE for allowing increases in carbon emissions and pollutants. They asserted that the new regulations would accelerate climate change and exacerbate asthma attacks, pulmonary, and other health problems (Heikkinen, 2018). After the public comment period had concluded, attorneys general (AGs) for 22 states and seven cities filed suit in the U.S. Court of Appeals to stop the proposal. New York State attorney general Letitia James said the Trump administration had replaced the Clean Power Plan with "this Dirty Power rule," which placed no cap on GHG emissions, undercut state and local efforts to require clean energy generation, and placed no cap on the chief driver of climate change, GHG emissions.

Initially, the COVID-19 pandemic had a benign effect in the area of air pollution. Manufacturing plants shuttered in many places, and traffic in cities across the land fell dramatically, as people were homebound to avoid exposure to the virus (Plumer & Popovich, 2020). However, Harvard researchers linked a slight increase in exposure to fine particulates to sharply higher death rates during the pandemic.

Biden administration changes. A federal appellate court invalidated ACE before Joe Biden was inaugurated in January 2021. This appeared to give the Biden administration carte blanche to create a new rule targeting GHG emissions from power plants (Farah, 2021).

The new administration was pragmatic in its approach to clean energy objectives in its first year. For example, when the criminal gang DarkSide's ransomware attack disrupted Colonial Pipeline's fuel delivery systems, President Biden waived CAA rules to avoid gasoline shortages on the East Coast (Soraghan, 2021). Nevertheless, evidence in support of stronger Clean Air Act regulations continued to be released. Research studies in 2021 showed "consistent, positive associations" between deaths from heart attacks and strokes and exposure to PM 2.5 at levels lower than those enforced in the Trump administration. Former EPA administrator Andrew Wheeler said in 2020 that agency career staff who sought to tighten the benchmark standard lacked conclusive evidence, but one year later that observation no longer held (Reilly, 2021b).

While the evidence supporting increased regulation of air pollutants continued to pile up, however, changes in Supreme Court membership moved it to the right, making it less supportive of regulatory action without what it considered a firm constitutional basis. Days before President Biden attended the global climate summit in Scotland, the court agreed to hear appeals from coal companies and Republican-led states seeking to limit the EPA's power to regulate carbon emissions under the CAA. A brief filed by coal state West Virginia and a dozen other states asked for immediate court action: "How we respond to climate change is a pressing issue . . . (but) Continued uncertainty over the scope of E.P.A.'s authority will impose costs we can never recoup" (Liptak, 2021). (In 2015, the investor-owned power companies opposed the CPP, but in 2022, utilities did not support the Republican-led fight against EPA's climate authority [Chemnick et al., 2022].)

At the end of 2021, the White House released the Unified Agenda, indicating the pace at which the president was addressing campaign pledges. In contrast to slow action in areas such as increased royalty rates that oil companies pay and endangered species protections, the EPA planned to move ahead by mid-2022 on limiting GHG emissions from coal-fired power plants (Brugger, 2021).

In early 2022, the EPA planned greater restrictions on power plant pollution, pointing to surges in emissions of sulfur dioxide, carbon dioxide, and other common pollutants (Reilly, 2022a). Yet when hearing oral arguments in West Virginia v. EPA (the case in which red states and coal companies asked that the agency be sharply limited in its ability to slash carbon dioxide under the CAA), most justices seemed prepared to limit EPA's oversight of power plants (Farah, 2022). Nevertheless, the agency pushed ahead in a new "good neighbor" plan to curb cross-state air pollution. It extended requirements to cut smog-forming emissions (such as ozone) to 26 states and industries beyond power plants (Reilly, 2022b).

FURTHER READING

American Geosciences Institute (AGI). "Air quality impacts of oil and gas," Petroleum and the Environment, 2018. https://www.americangeosciences .org/geoscience-currents/air-quality-impacts-oil-and-gas

American Lung Association. *Reports on "Key Findings," "Ozone Pollution," "Year-Round Particle Pollution," "Short-Term Particle Pollution," and*

"Seven Threats to the Nation's Air Quality," 2018 Report. https://www .stateoftheair.org

Banerjee, Neela. "12 states sue the EPA over proposed power plant regulation," *Los Angeles Times*, August 4, 2014.

Bote, Joshua. "After years of improvement, US air quality has gotten worse since 2016, report suggests," *USA Today*, November 1, 2019.

Brugger, Kelsey. "Biden agenda advances air regs; mixed progress elsewhere," *E&E News*, December 13, 2021.

Chemnick, Jean, Lesley Clark and Pamela King. "Why utilities didn't join the Supreme Court case against EPA," *Energywire*, March 10, 2022.

Clay, Karen and Nicholas Muller. "Recent increases in air pollution: Evidence and implications for mortality," Working Paper 26381, National Bureau of Economic Research, October 2019. https://www.nber.org /system/files/working-papers/w26381/w26381.pdf

EPA. "Our Nation's Air 2019." https://gispub.epa.gov/air/trendsreport/2019

Farah, Niina. "Clean Air Act gets boost as court dumps Trump carbon rule," *E&E News*, January 20, 2021.

Farah, Niina. "Supreme Court seems ready to limit EPA power plant oversight," *Energywire*, March 1, 2022.

Friedman, Lisa and Brad Plumer. "E.P.A. announces repeal of major Obama-Era Carbon Emissions Rule," *New York Times*, October 9, 2017.

Heikkinen, Niina. "Tensions high at public hearing for climate rule," *E&E News*, October 1, 2018.

Hildreth, Kristen. "Capitol to Capitol/Oct. 16, 2017." National Conference of State Legislatures (NCSL), 2017. https://www.ncsl.org/ncsl-in-dc /publications-and

Hildreth, Kristen. "Our American states," The NCSL blog; Washington, DC: National Conference of State Legislatures (NCSL), 2018. https:// www.ncsl.org/blog/2018/08/22/epa-unveils

Kuckro, Rod. "The Clean Power Plan: Its impact (or not)," *E&E News*, December 2, 2019.

Liptak, Adam. "Supreme Court to hear case on E.P.A.'s power to limit carbon emissions," *New York Times*, October 29, 2021.

Plumer, Brad and Nadia Popovich. "Traffic and pollution plummet as U.S. cities shut down for coronavirus," *New York Times*, March 22, 2020.

Popovich, Nadia. "America's skies have gotten clearer, but millions still breathe unhealthy air," *New York Times*, June 19, 2019.

Reilly, Sean. "Racial, economic disparities to soot exposure persist," *E&E News*, July 31, 2020.

Reilly, Sean. "National Academies to probe EPA work behind emission limits," *E&E News*, April 27, 2021a.

Reilly, Sean. "New research adds to evidence for stronger soot rules," *E&E News*, September 30, 2021b.

Reilly, Sean. "EPA: Power plant emissions spiked last year," *Greenwire*, February 18, 2022a.

Reilly, Sean. "EPA expands reach of ozone regulations," *Greenwire*, March 11, 2022b.

Soraghan Mike. "Biden responds to pipeline hack with EPA fuel waiver," *E&E News*, May 11, 2021.

U.S. Department of Energy, Energy Information Administration (EIA). "Greenhouse gases," July 15, 2021. https://www.eia.gov/energyexplained/energy-and-the-environment/greenhouse-gases.php

U.S. Environmental Protection Administration (USEPA). "Air quality improves as America grows: Status and trends through 2018," 2019. https://gispub.epa.gov/air/trendsreport/2019

Q32. HAVE CAFE (CORPORATE AVERAGE FUEL ECONOMY) STANDARDS REDUCED CARBON DIOXIDE EMISSIONS?

Answer: Yes, until 2018 when the Trump administration loosened standards. In 2021, however, the Biden administration tightened them back to pre-Trump levels.

The Facts: Fuel efficiency standards have engaged attention of policymakers for nearly 50 years. CAFE is the acronym for Corporate Average Fuel Economy, a fuel efficiency standard developed by Congress in 1975. Its purpose was to lower the amount of energy (gasoline, diesel) consumed by cars and light trucks. The averages are fleet-wide and need to be achieved by every automobile manufacturer for its car and truck fleet, every year. Two regulatory authorities monitor CAFE standards. The National Highway Traffic and Safety Administration (NHTSA) sets and enforces the standards; the Environmental Protection Administration (EPA) calculates the fuel economy levels for auto firms and also sets GHG standards (NHTSA, 2021).

Importance of fuel economy. The oil supply and price shocks of the 1970s convinced federal authorities and Congress to mandate that new cars and trucks improve fuel efficiency. Under pressure of the auto industry, Congress lowered the level, but environmental NGOs fought back—and the struggle continues to the present. The most recent fuel economy mandates set by the Biden administration call for a 10 percent greater emissions improvement than the previous (Trump) administration for model years 2024 to

26 vehicles. Environmentalists were divided in response. The Center for Biodiversity climate transport director Dan Becker said mileage standards were "weak" and did little to "alleviate consumers' pain at the pump." On the other hand, an Environmental Law & Policy Center attorney, Ann Jaworski, commented that the new standard was "a massive and much needed improvement." Transportation secretary Pete Buttigieg noted, "Today's rule is going to save 234 billion gallons of fuel by 2050 and move us into a less dependent future" (Skibell, 2022).

For most Americans, saving money at the pump and lessening dependence on the volatile global oil/gas market are important reasons for fuel economy standards. Increasing energy sustainability and reducing climate change require a longer time horizon. As energy scholars Cassedy and Grossman point out, small incremental improvements in fleet gas mileage can make large differences. Increasing mpg from 18 to 20 "would mean an additional saving of 200,000 barrels (bbl) countrywide of refined petroleum per day . . . (and going up to) 26 mpg could save more than 1.5 million bbl/day" (Cassedy & Grossman, 2017).

Conflict. Controversy over the CAFE standards broke out in the late 2010s when the Trump administration proposed a rollback of Obama administration efficiency targets (which it had dubbed SAFE for Safer Affordable Fuel Efficiency). The Trump administration vehicle rule it proposed would increase GHG emissions while reducing the prices of cars and trucks. A *New York Times* investigation contended that the motivation behind the campaign to lower fuel efficiency standards was the surplus of fossil fuels in the United States, the need to increase demand for oil/gas products, and to increase fossil fuel dominance in America's energy portfolio (Tabuchi, 2018).

California's approach to fuel economy. Throughout this controversy, California claimed that it had authority to set tougher emissions standards than the federal government. It made this assertion because its air regulatory agency, the California Air Regulatory Board (CARB), had a waiver under the 1990 amendments to the Clean Air Act (CAA) for greenhouse gases. Twenty-four governors (including three Republicans) supported the state in its fight against the Trump administration's proposed easing of CAFE standards. All were members of the U.S. Climate Alliance, a bipartisan network adhering to the Paris Summit's climate goals (Joselow, 2019a). Together these states represented more than 35 percent of U.S. vehicle sales. Shortly thereafter, four large automakers—Ford, Honda, Volkswagen, and BMW—signed a fuel economy agreement with California. The companies agreed to improve fuel efficiency of cars and light trucks by 3.7 percent each year, reaching the fleet-wide mileage goal of 54.5 miles per gallon (mpg) under lab conditions by 2026 (Joselow, 2019c).

Free market advocates, however, urged the president to "press forward" with rolling back Obama-era clean car standards despite criticism from automakers. These groups included the American Energy Alliance, the Competitive Enterprise Institute, the Heartland Institute, and FreedomWorks (Joselow, 2019b). The U.S. Chamber of Commerce, which normally sided with Republican administrations, in this case opposed the rollback believing it would create "chaos and confusion throughout the country." The Chamber also criticized the administration for attempting to dismantle the One National Program for fuel economy, which sought to ensure there was just one set of clean car standards (Joselow, 2019d). (Historically, the auto industry has sought certainty in the regulatory process, essential for long-term business planning.)

The Trump administration rollback. In September 2019, the Trump administration challenged California's authority to set its own fuel efficiency standards. It even warned that the state might have violated federal law by signing an agreement with automakers. The argument was that federal energy legislation gave Congress alone the power to set new standards for vehicle emissions. During a California fundraising trip, the president tweeted: "The Trump Administration is revoking California's Federal Waiver on emissions in order to produce the less expensive cars for the consumer, while at the same time making the cars substantially SAFER" (Cama, 2019). Within a week, Trump's EPA administrator, Andrew Wheeler, threatened revocation of federal highway funds in retaliation for California's lack of compliance with the CAA. True to form, environmental NGOs such as the Sierra Club promised protracted legal action against the administration if it tried to follow through on its threats.

The second part of the SAFE Vehicles Rule was released in March 2020. The nearly 2,000-page rule proposed to weaken fuel economy and GHG standards set by the Obama administration in 2012. Wheeler remarked that the regulatory burden for industry under the proposed rule would be lightened considerably, lowering manufacturing costs. However, environmental NGOs complained about significant increases in GHG emissions from transportation, the largest source of carbon pollution. The Environmental Defense Fund (EDF) estimated that the rollback would produce an additional 1.5 billion metric tons of carbon pollution, 18,500 additional premature deaths, and 250,000 more asthma attacks (Joselow, 2020a).

When it became clear that Trump had lost the 2020 presidential election, automakers (particularly Ford) sought to develop a united front before the Biden administration assumed power (Joselow, 2020b). In the weeks before Biden took office, the inspector general (IG) for EPA determined the rollback violated federal regulatory processes, overlooked concerns of career staff members, and ignored impacts on vulnerable communities.

Overall, the report concluded, the SAFE Vehicles Rule "undercut 30 years of rulemaking precedent" (Skilbell, 2021a).

Biden administration changes. Early in his administration, Biden reversed two Trump administration actions pertaining to California's special role in setting fuel economy standards. First, NHTSA announced that the federal government's authority to set vehicle emissions standards did not preempt that of California (Reilly & Bogardus, 2021). This meant that federal action represented a floor but not a ceiling for the states. In other words, states had to follow the minimum standards set by the federal government but could always make those standards more ambitious if they wished to do so. Second, the Biden EPA announced that there had been "significant issues" in the Trump administration's withdrawal of California's CAA waiver, and it would be reinstating the state's authority to set its own tougher tailpipe emissions standards for light-duty cars and trucks (Reilly, 2021).

California became a critical part of the Biden strategy to significantly reduce GHG emissions to address climate warming. Indeed, climate change was mentioned more frequently in the president's inaugural address than by any of his predecessors (Brugger, 2021a). It set the stage for early action on a strong role for clean energy in the administration. One part of that plan would increase average fuel economy of new passenger cars and trucks (for a total of 25 percent by model year 2026); a second part would increase electric vehicle (EV) sales by 50 percent by 2030. The new EPA administrator, Michael Regan, allocated a significant portion of his schedule to standards setting work, meeting with a range of auto company executives (Brugger, 2021b).

The Biden administration received support from 21 state AGs, the District of Columbia, and six cities in tightening fuel efficiency standards and GHG emission reductions. Their letter to EPA indicated the importance of the transportation sector to the realm of regulatory standards, as it represents 29 percent of total GHG releases. In its comments to the agency, the National Association of Clean Air Agencies (NACAA) pointed out: "Since the emission standards in the 2012 rule were adopted, clean vehicle technology and performance have progressed significantly, far more than anticipated and at lower cost, making a strengthening of those standards in this rulemaking feasible" (Skilbell, 2021).

FURTHER READING

Brugger, Kelsey. "Cheers, jeers as Biden skips the fine print," *E&E News*, April 29, 2021a.

Brugger, Kelsey. "Biden admin allows states to set stricter car rules," *E&E News*, December 22, 2021b.

Cama, Timothy. "Trump says he's revoking Calif.'s tailpipe waiver," *E&E News*, September 18, 2019.

Cassedy, Edward S. and Peter Z. Grossman. *Introduction to Energy: Resources, Technology, and Society*, 3rd ed. New York: Cambridge University Press, 2017.

Joselow, Maxine. "24 governors back Calif. over Trump," *E&E News*, July 9, 2019a.

Joselow, Maxine. "Free-market groups to Trump: Don't back down," *E&E News*, July 11, 2019b.

Joselow, Maxine. "Automakers buck Trump, sign fuel economy deal with Calif.," *E&E News*, July 25, 2019c.

Joselow, Maxine. "U.S. Chamber splits with Trump over car rules," *E&E News*, August 30, 2019d.

Joselow, Maxine. "Battle lines drawn as Trump finalizes rollback," *E&E News*, March 31, 2020a.

Joselow, Maxine. "Ford urges support for Calif. emissions deal," *E&E News*, December 1, 2020b.

Reilly, Sean. "EPA to restore Calif. authority over tailpipe emissions," *E&E News*, April 26, 2021.

Reilly, Sean and Kevin Bogardus. "Biden admin pushes plan to restore Calif. car rules waiver," *E&E News*, April 22, 2021.

Skilbell, Arianna. "Watchdog: Trump's rollback undercut EPA staff," *E&E News*, April 20, 2021a.

Skilbell, Arianna. "21 state AGs urge EPA to tighten clean car rule," *E&E News*, September 28, 2021b.

Skilbell, Arianna. "Biden unveils strongest fuel efficiency rule yet," *Greenwire*, April 1, 2022.

Tabuchi, Hiroko. "The oil industry's covert campaign to rewrite American car emissions rules," *New York Times*, December 13, 2018.

U.S. Department of Transportation, National Highway Transportation Safety Administration (NHSTA). "Corporate Average Fuel Economy (CAFE) standards," September 13, 2021. https://www.nhtsa.gov/laws -regulations/corporate-average-fuel-economy

Q33. DO MOST AMERICANS BELIEVE THAT CONTINUED CONSUMPTION OF FOSSIL FUELS WILL INCREASE THE SEVERITY OF CLIMATE CHANGE?

Answer: Yes, based on public opinion surveys; however, efforts to mitigate climate change face strong obstacles of extreme partisan resistance from the Republican Party, as well as opposition from a fossil fuel industry that realizes billions of dollars in profits annually from oil and coal consumption.

The Facts: In October 2018, the UN's Intergovernmental Panel on Climate Change (IPCC) warned that the world needed to do more about carbon emissions responsible for climate warming or face dire consequences. Asked about the report in a CBS interview, Trump dismissed it; "I think there's probably a difference [in climate]. But I don't know that it's man-made. . . . I will say this. I don't want to give trillions and trillions of dollars. I don't want to lose millions and millions of jobs" (Quinones, 2018; see Q27). At this point, fossil fuel industry managers and Republican Party leaders shrugged at the IPCC report, even as Democrats in Congress and environmental groups around the world declared that the report showed the need for immediate and decisive action to address the threat of climate change.

Public opinion surveys on climate warming, 2018–21. An Associated Press (AP) and National Opinion Research Center (NORC) survey released in September 2019 found that both Republicans and Democrats believed that climate change was happening, but they disagreed on its causes. Most Democrats said human activity was at least mostly responsible for climate change, while Republicans were more inclined to say that natural changes had some responsibility. As to the reality of climate warming, 76 percent of Americans thought it was real (this question was not broken down into partisanship), 9 percent said it was not happening, and 15 percent weren't sure. Reasons offered by the 9 percent who were skeptical included: distrust of the scientific evidence (55 percent), belief that the problem has been exaggerated (53 percent), and that it's a hoax (35 percent).

The poll also found stark partisan differences on energy questions related to climate change. Democrats favored research into renewables to a greater extent than Republicans; Republicans were positive about carbon-based fossil fuels, while Democrats saw little advantage to them. Some 67 percent of Republicans approved Trump's handling of climate change, compared to only 7 percent of Democrats and 29 percent of independents (AP/NORC, 2019). A few months later, the Pew Research Center reported that since conducting a 2016 poll, public support had increased for every type of renewable energy source and had declined for every source of fossil fuel energy, including offshore drilling and hydraulic fracturing.

The Pew survey found significant divisions among Republican voters. The millennials and Generation Z category (those younger than 30) held environmental (including climate change) views approximately like those of the general public. But conservatives, men, and older

Republicans (the majority of Republican voters) were less concerned about climate change (Funk & Hefferon, 2019).

A notable survey was undertaken by Colorado College's State of the Rockies project, which polled voters in eight of the Western states. The poll's focus was on energy in the West, and an overwhelming majority of respondents (nearly 90 percent) said they would require oil and gas companies to use the most modern and updated equipment to reduce emissions of methane, a powerful greenhouse gas. Only a quarter of respondents said climate change "has been greatly exaggerated" (Streater, 2020). The growing sense of urgency on the part of the public on the issue of climate change contrasts sharply with the sluggish and indifferent response from many lawmakers and industries (Stern & Dietz, 2020).

Another authoritative poll sponsored by the University of Chicago's Energy Policy Institute and AP/NORC compared public opinion change from 2018 to 2021 and found increased awareness and support for addressing the climate crisis. Sixty percent of respondents believed climate change was occurring. A small majority of respondents indicated they were willing to pay for solutions such as carbon pricing and modest fees to clean up power plants, increase EV use, and help fossil fuel-dependent communities change their economies. As expected, however, the poll exposed a wide partisan gulf in climate views. Only one-third of Republicans thought concerns about climate change were based on good science, compared to two-thirds of Democrats (Cusik, 2021).

In 2021, a Pew study found that Americans differed by race and ethnicity in their evaluation of climate proposals. About 7 in 10 Black adults (68 percent) and more than half of Hispanic adults (55 percent) identified assistance for lower-income communities as a top consideration in climate policy, compared to 38 percent of Caucasian adults. Black and Hispanic Americans were also more likely than Whites to say their own communities experienced environmental problems, such as air and water pollution and drinking water safety (Funk, 2021)

In mid-May 2021, the International Energy Agency (IEA) released a study indicating it was possible to achieve net-zero emissions by 2050—but doing so would require a complete and immediate energy system change. As IEA executive director Fatih Birol remarked, the scale and speed required to limit climate warming to no more than 1.5 degrees C. "make this perhaps the greatest challenge humankind has ever faced" (Iaconangelo & Richards, 2021). Supporters of climate action called the report a "turning point," while the American Petroleum Institute opined that policy leaders needed to be cautious so that demand for traditional fuels did not outstrip supply and accelerate "energy poverty." In 2022, a comprehensive UN climate report summarized years of studies from climate scientists, social scientists, energy experts; the report concluded that tackling climate change required a fundamental change of nearly everything (Harvey, 2022). A final Pew research report based on surveys done after the Russian invasion of the Ukraine and very high rates of inflation indicated that most Americans were not ready for such a dramatic change. More than two-thirds of respondents said the United States should keep a mix of fossil fuels and renewable energy; only 31 percent would support a phaseout of coal, oil, and natural gas (Bond, 2022).

Moderation in approach of the oil and gas industry? In early January 2020, the American Petroleum Institute (API) announced that it was spending at least $1 million (a pittance given the enormous resources of the oil/gas industry) to advertise its role in the fight against climate change. Its "Energy for Progress" effort would highlight benefits of oil and gas production in key swing states, such as Ohio, Colorado, and Pennsylvania. Also, the trade association would tell the public what a ban on fracking would cost—more than 7 million jobs and increased energy costs to consumers of $900 million. As part of the effort, the industry would call itself "gas and oil," putting the lower-carbon fuel in front (Cama, 2020). An industry coalition, the Oil and Gas Climate Initiative (OGGI), announced it would lower emissions of aggregated oil and gas operations by as much as 13 percent from 2017 levels. Critics said the target allowed increases in emissions overall (Anchondo, 2020), suggesting the slippery nature of such promises. Yet growing pressure from activist investors focusing on ESG (environment, social, and governance [Richards, 2020]) has alerted the industry to the increased necessity of self-regulation.

Meanwhile, Texas Oil and Gas Association president Todd Staples stated that his association needed to contribute to a solution to climate warming by reducing GHG output (Lee, 2020). Finally, the U.S. Chamber of Commerce's CEO, Tom Donohue, claimed that his organization, once diametrically opposed to policies to combat climate change, was now seeking bipartisan action on climate change (Sobczyk, 2020).

These efforts to reduce fossil fuel emissions, however, are countered by continued major private investments in oil, gas, and coal businesses. From 2010 to 2021, the private equity industry has pumped more than $1 trillion into the energy sector, and only about 12 percent of this amount has entered the coffers of renewable energy firms. More problematic is the secretive nature of this investment, which may result in private equity firms (such as Hilcorp, backed by equity giant Carlyle) purchasing assets of ConocoPhillips in Colorado and New Mexico and BP's Alaska operations to become the nation's largest emitter of methane (Yabuchi, 2021).

All companies in the United States—private or public—must follow environmental regulations, but private equity firms are exempt from a number of public financial disclosure rules. Publicly traded firms such as BP have a public relations incentive to divest from fossil fuel or other holdings contributing to climate change, but private firms such as Hilcorp backed by huge private equity firms (Carlyle) operate mostly out of the public eye. They can't easily be held accountable to the public.

Biden administration actions and policies. As mentioned in previous analyses, President Biden has been the most aggressive president in asserting the need for the United States to address the climate crisis. One such signal to the world was Biden's decision to have the United States rejoin the Paris climate accord, while at home seeking to overturn many of the Trump administration's environmental rollbacks. Biden's appointments increased the staff dedicated to climate action, as seen particularly in appointments to Energy, Interior, and the EPA and his decision to bring on both John Kerry (former secretary of state and climate ambassador) and Gina McCarthy (former EPA administrator) as dedicated White House advisors on climate issues. The administration promised billions in projects to improve resilience of the electricity grid, and to pare back federal support for fossil fuels. The Biden administration's announced agenda for clean energy also went further than previous presidencies in several areas but particularly in wind, as the administration announced a plan to develop large-scale wind farms along almost the entire coastline of the United States (Clark, 2021).

Initially, the ambitious, several trillion-dollar infrastructure proposals would have included billions in rewards to utilities for switching from burning fossil fuels to renewable energy sources, as well as numerous federal grants and subsidies to support electric vehicles and renewables. However, the size of the infrastructure proposal—including significant green energy elements—was reduced in order to gain the support of conservative and moderate Democrats in Congress (Davenport, 2021).

At a news conference marking his first year in office, President Biden said he supported breaking up or reducing significantly in size the "Build Back Better Act," emphasizing that the $555 billion in climate spending was an area of agreement: "I think we can break the package up, get as much as we can now and come back and fight for the rest later (Sobczyk, 2022). Administration strategists sought to use the regulatory means to make definable progress on climate goals, but without a clear majority in Congress, even the strongest of regulations would not withstand court scrutiny. That close examination was likely on an upcoming case concerning the authority of the EPA.

Coal companies and Republican-led states have asked the Supreme Court to severely limit the EPA's latitude to craft rules under a new version of the 2015 Clean Power Plan, alleging that agency discretion would lead to a destabilizing restructuring of America's energy system. Challengers seek to constrain EPA from recommending that states employ a "best system of emission reduction" (including using more renewable energy generation, emissions trading, or a carbon rule). Biden's nominee (Judge Ketanji Brown Jackson) to the Supreme Court faced both sides of this issue when she addressed the Senate Judiciary Committee. As a circuit court judge, she observed that "political questions" were outside the jurisdiction of the high court (King, 2022a).

In a decision expected in summer 2022, the high court will consider whether Biden administration proposals violate the "major questions" doctrine that courts should not defer to regulatory agencies on issues of great political or economic significance. Republican state attorneys general and conservative groups have challenged EPA's fuel economy rule that bolsters major climate regulations. They believe that EPA cannot impose rules that would force an industry to employ a new technology and move toward renewables (King, 2022b).

FURTHER READING

Anchondo, Carlos. "Big Oil CO2 announcement riles critics," *E&E News*, July 17, 2020.

AP/NORC. "The politics of climate change." https://apnews.com/82e8e6fd7bd43436cbf5208ee1558d6b1

Bogel-Burroughs, Nicholas and Carol Davenport. "Climate protesters snarl traffic but Washington still goes to work," *New York Times*, September 23, 2019.

Bond, Camille. "Report: Most Americans don't back 100% renewables," *Energywire*, March 2, 2022.

Cama, Timothy. "Oil and gas industry launches climate campaign," *E&E News*, January 7, 2020.

Clark, Lesley. "Gina McCarthy: White House to use 'regulatory authority' on climate," *E&E News*, July 14, 2021.

Cusik, Daniel. "Poll shows public shift on clean energy, EVs, climate," *E&E News*, October 26, 2021.

Davenport, Coral. "Key to Biden's climate agenda likely to be cut because of Manchin opposition," *New York Times*, October 15, 2021.

Funk, Cary. "Key findings: How Americans' attitudes about climate change differ by generation, party and other factors," May 26, 2021. https://www

.pewresearch.org/fact-tank/2021/05/26/key-findings-how-americans
-attitudes-about-climate-change-differ-by-generation-party-and-other
-factors/

Funk, Cary and Meg Hefferon. "U.S. public views on climate and energy,"
November 25, 2019. https://www.pewresearch.org/science/2019/11/25/u-s
-public-views-on-climate-and-energy/

Harvey, Chelsea. "IPCC: For climate goals, it's 'now or never'," *Greenwire*,
April 4, 2022.

Heberlein, Thomas. *Navigating Environmental Attitudes*. New York: Oxford
University Press, 2012.

Iaconangelo, David and Heather Richards. "Groundbreaking climate
report rattles EV, gas," *E&E News*, May 19, 2021.

King, Pamela. "Jackson questioned on key doctrine in climate litigation,"
Greenwire, March 23, 2022a.

King, Pamela. "Inside a legal doctrine that could derail Biden climate regs,"
Greenwire, April 11, 2022b.

Lee, Mike. "Texas oil association shifts on climate," *E&E News*, January 15, 2020.

Quinones, Manuel. "As Dems slam Trump's comments, most Republicans
shrug," *E&E News*, October 15, 2018

Richards, Heather. "These 3 letters could end fossil fuels—or green wash
them," *E&E News*, June 15, 2020.

Sobczyk, Nick. "U.S. Chamber CEO calls for bipartisanship on climate,"
E&E News, January 9, 2020.

Sobczyk, Nick. "Biden resets BBB: 'I think we can break the package up,'"
E&E News, January 20, 2022.

Stern, Paul C. and Thomas Dietz. "A broader social science research
agenda on sustainability: Nongovernmental influences on climate
footprints," *Energy Research and Social Science*, Vol. 60 (2020):
101401.

Streater, Scott. "Westerners want more climate action, less drilling,"
E&E News, February 20, 2020. See: "Energy in the West." https://www
.coloradocollege.edu/other/stateofthe

Tyson, Alec. "On climate change, Republicans are open to some policy
approaches, even as they assign the issue low priority," Pew Research
Center, July 13, 2021. https://www.pewresearch.org/fact-tank/2021/07/23
/on-climate-change-republicans-are-open-to-some-policy-approaches
-even-as-they-assign-the-issue-low-priority/

Yabuchi, Hiroko. "Private equity funds, sensing profit in tumult, are
propping up oil," *New York Times*, October 13, 2021.

Q34. HAVE CLEAN WATER LAWS AND REGULATIONS IMPROVED U.S. WATER QUALITY?

Answer: Yes.

The Facts: Water quality is another issue area expressing conflict between public health (and other environmental values) and economic development. In the United States, there are two categories of water for personal use—groundwater pumped to the surface and used by about 40 percent of the population for drinking, bathing, and cooking; and surface water delivered to people's homes (the rest of the population) from oceans, lakes, rivers, and streams.

"Water pollution" occurs when harmful substances, often chemicals or microorganisms, contaminate a river, stream, lake, pond, ocean, aquifer, or some other water body. Such pollution can have negative consequences for both public and ecosystem health (Denchak, 2018). Estimates vary, but roughly one-half of water pollution in the United States originates from industrial activities, for example, dumping of toxic industrial wastes, mining, accidental oil leakages, and burning of fossil fuels, among other sources. Specialized terms such as "produced water" describe what happens when the extraction of oil and gas creates a massive waste stream of chemical-laced, saline, and often radioactive wastewater. The stream comes to the surface with the oil and gas; it is a "formation fluid" and often is high in naturally occurring contaminants and chemical additives, for example, hydraulic fracturing fluid, acids, and other chemical products used in routine well operations (Grinberg, 2019).

Condition of U.S. waters. In the last half-century, presidential administrations, Congress, and government agencies established a relatively comprehensive legal and regulatory system to protect waters. Q28 described two of the three primary laws, the Clean Water Act (CWA) of 1972 and the Waters of the United States (WOTUS). For the third law, the Safe Drinking Water Act (SDWA) of 1974, the EPA developed lists of water contaminants posing a potential danger to public health. Research was done on the contaminants, and a determination made of the maximum level of contaminant that could be permitted in drinking water. The primary and secondary types of contaminants regulated were disinfectants, disinfectant byproducts, inorganic and organic chemicals, and microorganisms, including bacteria, viruses, and protozoa—all of which caused illnesses if consumed by humans (EPA, 2017).

These provisions did not apply to hydraulic fracturing (HF) because of the "Halliburton loophole," which exempted HF from the Underground Injection Control (UIC) provisions of the SDWA (the UIC part of the Safe Drinking Water Act governs underground storage of contaminants and regulates wells used to pressure oil and gas to the surface). Halliburton was a huge oil services firm that sought the exemption in the 1990s when its CEO was Dick Cheney, who later became George W. Bush's vice president, playing a key role in the development of the 2005 energy bill that contained the Halliburton loophole (McBeath, 2016). The exception prevented the EPA from regulating HF, leaving hydraulic fracturing to pro-development state agencies—which industry preferred.

As with the terms of the SDWA, the federal government under the Clean Water Act also shares responsibility with the states in the protection of water quality. States are primary agents for managing diffuse (nonpoint) sources of water pollution, for example, from construction sites and farms (a central source of pollution as pesticide/herbicide residues flow into drainage ditches and streams). EPA is the primary regulator, however, of pollution from point sources such as municipal storm water runoff, industrial plants, and wastewater plants (GAO, 2016).

Since the onset of national environmental laws and regulations in the late 1960s and 1970s, contamination of U.S. waters has been reduced significantly. Still, critics lament that over one-third of Americans lack ready access to safe and healthful water. They allege that currently unregulated contaminants are harmful to public health; that EPA has not added any toxic chemicals to the SDWA list; state governments have not monitored public water systems; and that EPA has not prioritized vulnerable populations—such as infants, children, pregnant women, and elderly individuals with a history of serious illnesses—when reviewing drinking water regulations (EWG, 2019).

Obama administration clean water policy. The Obama administration's Waters of the United States (WOTUS) rule stemmed from two Supreme Court decisions (in 2001 and 2006) that left it unclear whether small bodies of water (streams, creeks, wetlands, etc.) not directly and continuously connected to large waterways, such as the Mississippi River, Chesapeake Bay, or Puget Sound, were federally protected under the Clean Water Act.

Under the Obama administration's proposed rule, all waterways with a streambed, two banks, and a high-water mark would have CWA protection, as would wetlands during dry seasons and ditches/gulches wet only during snowmelt or flooding. The rule would apply to 60 percent of the nation's water bodies, and while it was praised by environmental organizations, it encountered stiff opposition from farmers, ranchers, and the oil and gas industry, among others. For example, farmers would need EPA permits to spread chemical fertilizers and pesticides/herbicides that might run off into adjacent water bodies. The American Farm Bureau and the U.S. Chamber of Commerce challenged the rule, as did governors of 27 states (Davenport, 2015). A legal battle over WOTUS raged from 2015 to 2019, at which point the Trump administration replaced WOTUS with its own rule.

Trump administration clean water policy. In the 2016 election contest, Donald Trump pledged to restore property of farmers, ranchers, and oil/gas companies, compromised by "federal overreach," as part of the campaign to "Make America Great Again." The Trump administration subsequently proposed dramatic rollbacks of clean water regulations, including changes to wastewater permitting processes regarding oil and gas development and toxic contamination standards. The Trump administration also announced a new Navigable Waters Protection Rule (NWPR) that removed federal oversight from 51 percent of wetlands and 18 percent of streams across the United States. It protected streams flowing to traditional navigable waters, but excluded wetlands intersected by "ephemeral" streams and most ditches (Wittenberg, 2019b). The rule was finalized but immediately challenged in court, and as of early 2022 the legal jousting over NWPR was ongoing. The Biden administration has moved cautiously on the WOTUS rule because of uncertainty over courts' treatment of agency rules when power between legislative and executive branches is divided (Northey, 2022a).

The drilling industry had dealt with huge flows of contaminated ("produced") waters by injecting them into the ground, but this practice became problematic on several fronts. The practice was linked to Oklahoma earthquakes, and in New Mexico and Texas, storage capacity of underground wells reached capacity. The industry proposed to use this wastewater for beneficial purposes, for example, irrigating crops, watering livestock, municipal use, or letting it flow into rivers and streams. Some oil states had sought EPA approval for an NPDES permit under the CWA to allow such discharges, which prompted the Supreme Court decision Maui County v. Hawai'i Wildlife Fund. In that 2019 ruling, the high court crafted a new standard that CWA permits were required when "there is a direct discharge from a point source into navigable waters or when there is the *functional equivalent of a direct discharge*" (King & Marshall, 2020). Environmental NGOs such as the League of Conservation Voters called this treatment of produced water the "dirty water rule" and opined that it would "put clean drinking water for tens of millions of people at risk, especially the low-income communities and communities of color already disproportionately impacted by polluted water" (Milman, 2020).

The Safe Drinking Water Act has not been updated since 1998, an important issue for environmentalists specializing in water security. The Environmental Working Group (EWG) sought regulation of perchlorate, a rocket fuel ingredient linked to thyroid problems, but the Trump EPA's proposed rule allowed concentrations in drinking water 10 times higher than set by states. An EWG science adviser objected: "The science on perchlorate is very clear: It harms infants and the developing fetus. . . . (EPA's lack of action) will endanger the health of future generations of kids" (Wittenberg, 2019a). (In this case, the Trump administration decision to not issue a drinking water rule for perchlorate was upheld by the Biden administration, to the consternation of environmental groups. The EPA said in early 2022 that the earlier decision had been based on the best available peer-reviewed science, and the agency had taken adequate steps to reduce exposure to the chemical [Northey, 2022b].)

The last chemicals regulated by SDWA are toxic fluorinated substances called PFAS contaminating most supplies of drinking water in the United States. In this case, the Democratic-controlled House of Representatives added mandates for EPA to establish PFAS standards and cleanup require-ments to a "must-pass" Defense Department appropriations bill in 2019, but these were stripped from the bill at the insistence of Trump and Senate Republicans. In 2019, President Trump said: "From day one, my adminis-tration has made it a top priority to ensure that America has among the very cleanest air and cleanest water on the planet. We want the cleanest air, we want crystal clean water, and that's what we're doing and what we're working on so hard." Critics scoffed at Trump's rhetoric, accusing him of 1) taking credit for pollution reductions largely the result of efforts of his predecessors, and 2) engaging in an aggressive effort to weaken vital air and water pollution rules (Guillen et al., 2019).

Biden administration water policy changes. In his first day in office, President Biden addressed contentious rules and policies concerning water quality: wetlands protections, permitting oil and gas company field opera-tions, assessments and monitoring of toxic chemicals and toxic waste sites. Also under review were regulations pertaining to lead and copper water pipes, the streamlining of Corps of Engineers rules for oil and gas pipelines, and states' abilities to block energy projects under provisions of the CWA. The chief of staff issued a memorandum freezing all midnight (late-term) regulations imposed by the previous administration (Crunden & Northey, 2021). Incoming EPA administrator Michael Regan established an EPA Council on PFAS, to develop a strategy to address the chemical contami-nants on land and in air and water. Before the end of the year, the admin-istrator announced that the agency would require manufacturers to test and report publicly the amount of "forever" chemicals contained in nonstick pans, stain-resistant furniture, and waterproof clothing (Friedman, 2021). However, the EPA definition of the number of compounds to be tested and regulated was narrower than critics alleged was necessary, and a watchdog organization sued the agency (Crunden, 2022). (See also discussion of problems in regulatory authority of the FDA and EPA concerning chemicals in food packaging [Crunden & Wittenberg, 2022].)

One of the late-term Trump regulations repealed early in the Biden administration was the CWA Section 401 Certification Rule, which reduced the amount of time that states had to deny approval of proposed energy projects from one year to 60 days. The override provided states and Native American tribes more time to negotiate difficult energy projects (Northey, 2021). The Trump-era WOTUS rule was under court appeal when Biden assumed the presidency. By late August 2021, Judge Rosemary Marquez of the Arizona U.S. District Court said the Trump rule was too flawed to stay in place, and it was vacated nationally. Because the Obama WOTUS regulation had been rescinded by the Trump rule, this left in force a 2008 guidance of the George W. Bush administration that deferred to a 1986 regulation. In the absence of a new act of Congress, an alternative was that the Obama 2015 rule might be revived (Northey & King, 2021).

The Biden administration's plans for vastly expanded water infrastructure spending also reflected clean water concerns raised by a number of drinking water crises that had garnered headlines in the previous years. In 2014, for example, poor majority-Black communities in the city of Flint, Michigan, experienced a water crisis when a change in water source left residents drinking water contaminated with lead from old pipes after being told it was safe. Four years later, officials in Benton Harbor, Michigan, were told their water was contaminated by lead, but word did not reach all residents until later, and the problem is unresolved (Smith, 2021). Many American communities, and disproportionately those of poor and minority residents, are similarly afflicted.

FURTHER READING

Anchondo, Carlos and Lesley Clark. "Trump issues long-awaited fracking ban report," *E&E News*, January 15, 2021.

Crunden, E. A. "Trump leaves murky Superfund legacy," *E&E News*, January 13, 2021.

Crunden, E. A. "Suit targets EPA's 'forever chemicals' approach," *Greenwire*, May 2, 2022.

Crunden, E. A. and Ariel Wittenberg. "Inside FDA's 'forever chemicals' catastrophe," *Greenwire*, March 7, 2022.

Crunden, E. A. and Hannah Northey. "Water, chemical rules on chopping block," *E&E News*, January 20, 2021.

Davenport, Coral. "Obama announces new rule limiting water pollution," *New York Times*, May 27, 2015.

Denchak, Melissa. "Water pollution: Everything you need to know," Natural Resources Defense Council, May, 14, 2018. https://www.yourearth.net/water-pollution-everything-you-need-to-know

Environment Working Group (EWG). "Fixing our nation's drinking water policy," updated 2019. https://www.ewg.org/tapwater/fixing-drinking-water-policy.php

Friedman, Lisa. "A move to rein in cancer-causing 'forever chemicals,'" *New York Times*, October 18, 2021.

Grinberg, Andrew. "We all live downstream," The Clean Water Blog, March 28, 2019. https://www.cleanwateraction.org/blog/1655?field_issues_tid=69&field_region_state_tid=All

Guillen, Alex, Annie Snider, and Eric Wolff. "Fact Check: Trump's environmental rhetoric versus his record," *Politico*, July 8, 2019.

King, Pamela and James Marshall. "Clean Water Act ruling tees up coal ash brawl," *E&E News*, April 28, 2020.

McBeath, Jerry A. *Big Oil in the United States: Industry Influences on Institutions, Policy, and Politics*. Denver: Praeger, 2016.

Milman, Oliver. "Trump administration strips pollution safeguards from drinking water sources," *The Guardian*, January 29, 2020.

Northey, Hannah. "Biden moves to blunt Trump water permitting rule," *E&E News*, August 20, 2021.

Northey, Hannah. "EPA, Supreme Court jockey to define Clean Water Act's reach," *Greenwire*, January 31, 2022a.

Northey, Hannah. "EPA won't regulate rocket fuel in drinking water," *Greenwire*, April 1, 2022b.

Northey, Hannah and Pamela King. "What's next for WOTUS after judge jettisons Trump rule," *E&E News*, August 31, 2021.

Popovich, Nadia, Livia Albeck-Ripka, and Kendra Pierre-Louise. "95 environmental rules being rolled back under Trump," *New York Times*, December 21, 2019

Smith, Mitch. "More lead-tainted water in Michigan draws attention to nation's aging pipes," *New York Times*, October 16, 2021.

U.S. Congress, Governmental Accountability Office. "Water quality and protection." October 14, 2016. https://www.gao.gov/water-quality-and-protection

U.S. Environmental Protection Administration. "Drinking water regulations and contaminants," 2017. https://www.epa.gov/sdwa/drinking -water-regulations-and-contaminants

U.S. Environmental Protection Administration. "Overview of the navigable waters protection rule," 2020. https://www.epa.gov/sites/default /files/2020-01/documents/nwpr_fact_sheet_-_overview.pdf

Wittenberg, Ariel. "Health groups jeer EPA's 'stunningly high' perchlorate limit," *E&E News*, May 24, 2019a.

Wittenberg, Ariel. "Trump admin repeals Obama-era WOTUS rule," *E&E News*, September 12, 2019b.

Q35. HAVE OIL AND GAS EXPLORATION AND DEVELOPMENT REDUCED CRITICAL HABITAT OF ENDANGERED SPECIES IN THE UNITED STATES?

Answer: Yes, but without a significant deleterious impact on the listed species (yet).

The Facts: The number of species threatened with extinction as a result of human activity continues to multiply. The main law the United States has on the books to protect endangered species is the 1973 Endangered Species Act (ESA), which was actually passed into law with broad bipartisan support. The ESA has not been reauthorized since 1992, however, because Republican lawmakers, officials, and voters all came to see the law as resulting in excessive government interference with property rights and burdensome roadblocks to economic development. Only in a very small number of ESA cases, however, has government action actually "taken" property under the Fifth Amendment (Meltz, 2013). The Trump administration pledged to "reform" the ESA, and the impacts of the attempted rollback are still being felt by officials and agencies in the Biden administration.

The Endangered Species Act defines a species as endangered when it is "in danger of extinction throughout all or a significant portion of its range." Under ESA provisions, such species are automatically protected against harm, harassment, collection, or killing, with few exceptions. As a practical matter, this protection extends to habitat upon which the species depends for survival. A threatened species, on the other hand, is one that is "likely to become an endangered species within the foreseeable future" in the same range parameters. Section 4(d) of ESA specifies conditions applied to threatened species, based on their risk of extinction. Under ESA a rigorous evaluation process ascertains whether the species is overutilized for commercial, recreational, scientific, or educational purposes. The best scientific and commercial information available is the only basis for the listing (NOAA, 2019).

Critical habitat had long been defined as what is "essential for the conservation of the species," but the Trump administration challenged that definition. Late in Trump's term, Fish & Wildlife Service (FWS) scientists provided a "final" definition: "For the purpose of designating critical habitat only, habitat is the abiotic and biotic setting that currently or periodically contains the resources and conditions necessary to support one or more life processes of a species." Because the language made no mention of future areas, environmental NGOs criticized it for leaving unoccupied areas unprotected (Doyle, 2020b). Critical habitat had been designated for about 46 percent of the 1,500 endangered or threatened species (as of 2015).

The revised critical habitat definition advanced by the Trump administration permitted timber industry operations in Oregon, where resource development had been in conflict with preservation of the spotted owl, an endangered species, for decades.

ESA changes in the Obama administration. At the end of the George W. Bush administration, agencies such as FWS were given the latitude to decide whether potentially threatening activities (new dams, mines, logging) could proceed; too, they decided if the impact of oil/gas activities on climate change was considered. When President Obama reversed this practice, business groups objected, but environmentalists applauded (Goldenberg, 2009).

A second action of the Obama presidency was to find accommodations to the management plans of BLM and the Forest Service (FS) regarding threatened species, such as the sage grouse. The population of this iconic bird, found in 11 Western states, had been cut in half because of oil and gas development, ranching, and other economic activities. The administration negotiated with the states, the oil and gas industry, and other stakeholders to save the grouse and avoid an ESA listing—which would have placed more stringent limitations on economic development in sage grouse habitat (Streater, 2018). The pragmatic 2015 plan was supported by land managers (USDOI, 2015), environmental NGOs (such as the Audubon Society), but not by industry or ranchers.

Toward the end of the Obama administration, however, it made little progress in convincing the Republican majority in Congress to reauthorize ESA. The administration crafted rule changes to make the process more transparent, efficient, and collaborative. The proposals included increased rigor in petitions to list (or delist) species, change their status, or change habitat designations; limiting petitions to one species at a time (eliminating mega-petitions with 100+ species listed); submitting petitions to state fish and wildlife agencies in advance; and requiring petitioners to meet steeper requirements, such as online posting (Taylor & Hiar, 2015). For Republican members of Congress and most business interests, such proposals were palliatives. During Obama's two terms, 340 new species had been added to the endangered species list, and 29 species had been recovered enough to be removed from it.

Trump administration changes in endangered/threatened species policy. Having campaigned against federal overreach, the Trump administration reduced the amount of critical habitat needed to preserve endangered species (DOI, n.d.). BLM revised the Obama-era greater sage grouse conservation plans, removing nearly all of the 10 million acres labeled "sagebrush focal areas" and identified as critical habitat. Governors in affected Western states were ecstatic, but the action riled environmentalists (Streater, 2018). BLM also continued its leasing of lands for oil and natural gas exploration within designated wildlife corridors (Streater, 2019).

However, Chief Judge Brian Morris of the U.S. District Court in Montana ruled that planned oil/gas leasing in Wyoming intersected greater sage grouse habitat, and BLM was ordered to temporarily withdraw the tracts (Streater, 2020). In opposition to the judge's order, the Trump BLM issued "midnight" decisions (very late in the Trump administration) with "modifications, exemptions, and waivers" to the Obama mandates on protection of the sage grouse habitat. Analysts said they could be used to bolster campaigns of Republican candidates against Democrats in the 2022 and 2024 elections (Streater, 2021a).

The Trump administration did make important changes to the rules and regulations of species conservation under the ESA. First was the case of the American burying beetle. This inch-and-a-half-long beetle with orange markings on its wings (and a one-year life span) once inhabited 35 states but in 2020 was found only in 9. Since being declared endangered in 1989, protecting the beetle has interfered with land development, agriculture, transportation (the XL pipeline), and utility operations, alleged the Independent Petroleum Association of America (IPAA) among other interests. A new FWS rule down-listed it from endangered to threatened. This was related to a second change in implementing regulations, removing the "blanket" 4(d) rule (granting threatened species the same safeguards as endangered ones from harm), which environmentalists saw as weakening protections of vulnerable species such as butterflies and bird populations from climate change die-offs (Aton, 2019).

Challenges to ESA, and the record, 1980 to 2020. In 2019, the director of the Fish & Wildlife Service said that since Trump took office, the agency had completed or proposed delisting of 9 recovered domestic species. Some 25 domestic species had been proposed for or completed a down-listing from endangered to threatened. The Trump administration's proposals to reform ESA and emphasize recovery were advertised by agencies and the White House as major, bedrock change. Jake Li, director of biodiversity for the Environmental Policy Innovation Center, counted 33 specific changes made by Trump's Interior Department from 2017 through 2019, but 23 were minor or had a negligible impact (Doyle, 2020a).

The United States experienced a net loss of protected areas in the 2010s due to the Trump administration withdrawal of about 2 million acres from the Bears Ears and Grand Staircase Escalante national monuments—acres disappearing to roads, houses, pipelines, and other development (Lee-Ashley, 2019). (These actions subsequently were reversed by the Biden administration, which planned extensive work to shore up the sites [Yachnin, 2021].) In its first three years, the Trump administration proposed listing fewer than two dozen species as endangered/threatened, a dramatically smaller number than in previous administrations. During eight years of the Obama administration, some 360 species were listed as threatened or endangered. In the Clinton administration, 523 species were listed over eight years. Of the Republican presidents, in the Reagan administration, there were 253 listings; in the George H. W. Bush administration (of four years), 232 species were listed; and in the George W. Bush administration, 62 species (Green, 2019).

Biden administration changes in endangered/threatened species policy. The approach of the Biden administration to endangered species protection in considering both nonrenewable and renewable energy exploration and development appeared diametrically opposed to that of the Trump administration, but the processes followed were similar. First, it pledged a wholesale review and potential reversal of Trump administration actions pertaining to the Endangered Species Act. Biden's transition team also announced that the president would issue an executive order freezing the records of decision established by federal agencies that loosened protections of the greater sage grouse (Streater, 2021b). (Ultimately, it was a federal district court judge who invalidated 600 oil and gas leases in the species' locales [Farah, 2021] and the next year blocked five more Trump-era lease sales that were noncompliant [Farah, 2022].)

Unlike the Trump administration, Biden agencies emphasized wildlife protection in renewable as well as oil/gas energy zones. They proposed protections of plant species (a buckwheat flower) at a Nevada lithium plant site, as well as ESA protection for the lesser prairie chicken after voluntary efforts had failed. The Interior Department also restored protections to migratory bird species (specifically by reviving bans on unintentional killings) (Doyle, 2021). Then in June 2021, two major ESA watchdog agencies (FWS and NOAA) acted to revise, rescind, or reinstate five Trump-term regulations that conflicted with the objectives of the Biden-Harris administration, such as restoring critical habitat provisions (CRS, 2021). Environmentalists praised the Biden administration for "restoring critical protections for imperiled species," but critics claimed that the Biden White House was giving agencies power to "weaponize the ESA against rural Americans, bogging them down in years of litigation, burdensome regulations and government overreach" (Doyle, 2021a; Doyle, 2022).

A report on extinctions in 2021 noted that since adoption of ESA in 1973 it has "been successful at preventing the extinction of more than 99% of species listed." Altogether, 54 species had been delisted because of recovery, and another 56 species had been down-listed from endangered to threatened (Doyle, 2021b). These successes are all the more remarkable because they came during a period of decades in which U.S. fossil fuel output soared to top that of other nations, and renewables production increased markedly as well (Boylan et al., 2021).

FURTHER READING

Aton, Adam. "Trump admin rolls out rule changes to limit law's reach," *E&E News*, August 12, 2019.

Boylan, Brandon, Jerry McBeath, and Bo Wang. "Implementation deficits in endangered species protection: Comparing the U.S. and Chinese approaches." In *Imperiled: The Encyclopedia of Conservation*. Cheltenham, UK: Edward Elgar Publishing, 2021.

Carpenter, P.A. and P.C. Bishop. "The seventh mass extinction: Human-caused events contribute to a fatal consequence," *Future*, Vol. 41, no. 10 (December 2009): 715–22.

Doyle, Michael. "Trump claims ESA successes. Not so fast, critics say," *E&E News*, January 2, 2020a.

Doyle, Michael. "'Habitat' defines a heated new debate over protections," *E&E News*, December 1, 2020b.

Doyle, Michael. "Interior restores migratory bird protections," *E&E News*, September 29, 2021a.

Doyle, Michael. "Biden admin to uproot Trump 'critical habitat' policies," *E&E News*, October 26, 2021b.

Doyle, Michael. "Funding bill showcases species recovery," *Greenwire*, March 10, 2022.

Farah, Niina. "Judge nixes 600 oil and gas leases in sage grouse county," *E&E News*, June 11, 2021.

Farah, Niina. "Judge axes Trump-era lease sales to protect sage grouse," *Greenwire*, March 14, 2022.

Goldenberg, Suzanne. "Obama reverses Bush decision on Endangered Species Act," *The Guardian*, March 3, 2009.

Green, Miranda. "Rate of new endangered species listings declines under Trump," *The Hill*, December 5, 2019.

Lee-Ashley, Matt. "How much nature should America keep?" Center for American Progress, August 6, 2019. https://www.americanprogress.org/article/much-nature-america-keep

Meltz, Robert. "The Endangered Species Act (ESA) and claims of property rights 'takings,'" U.S. Congress, Congressional Research Service, January 7, 2013. https://fas.org/sgp/crs/misc/RL31796.pdf

Streater, Scott. "BLM's long-awaited revisions emphasize state 'flexibility,'" *E&E News*, December 6, 2018.

Streater, Scott. "Analysis: BLM continues leasing in wildlife corridors," *E&E News*, April 19, 2019.

Streater, Scott. "'Big win' for sage grouse as BLM cuts 247K acres from sale," *E&E News*, September 18, 2020.

Streater, Scott. "Interior finalizes revisions to Obama's sage grouse plans," *E&E News*, January 13, 2021a.

Streater, Scott. "Biden takes first step to ditch Trump sage grouse revisions," *E&E News*, January 20, 2021b.

Taylor, Phil and Corbin Hiar. "Obama overhaul draws GOP support, raises legal questions," *E&E News*, May 19, 2015.

U.S. Congress, Congressional Research Service (CRS). "Final Rules Amending ESA Critical Habitat Regulations," January 25, 2021. https://crsreports.congress.gov/product/pdf/IF/IF11740

U.S. Department of Commerce, National Ocean and Atmospheric Agency (NOAA). "What is the difference between a threatened and endangered species?" April 22, 2019. https://oceanscience_noaa.gov/facts/endangered.html

U.S. Department of the Interior (USDOI). "Historic conservation campaign protects greater sage-grouse," September 22, 2015. https://www.doi.gov/pressreleases/historic-conservation-campaign-protects-greater-sage-grouse

U.S. Department of the Interior (USDOI). "Threatened & endangered species active critical habitat," n.d. https://ecos.fws.gov/ecp/report/table/critical-habitat.html

Yachnin, Jennifer. "Biden is set to restore monuments. What happens next?" *E&E News*, October 8, 2021.

Q36. DID ENVIRONMENTAL NONGOVERNMENTAL ORGANIZATIONS (NGOs) FILE FEWER LAWSUITS AGAINST GOVERNMENT AGENCIES IN THE 2010s THAN THE 1970s?

Answer: No. Environmental litigation against federal agencies increases in conservative (and populist) Republican administrations, but there is no clear trend line.

The Facts: Roughly 20 million Americans are members of environmental nongovernmental organizations (NGOs), including well-known groups like the Sierra Club, Greenpeace, and the Environmental Defense Fund. These organizations typically engage in a wide range of activities to advance their environmental goals, including initiatives to educate the public, mobilizing citizens to action, lobbying policymakers, and pursuing legal remedies in the courts. During the past half-century, as NGO specialist Michael Kraft observes, U.S. environmental policy has evolved from federally dominant regulation (prompting lawsuits from NGOs), to concern for reform efforts emphasizing efficiency, to searching for the best way to attain the condition of sustainable development. A variety of policy tools have become popular, such as market-based incentives (MBIs) and public–private partnerships (Kraft, 2015). Three terms clarify activities of environmental NGOs when they come into conflict with federal agencies— usually because the environmental groups believe that agency policies or practices are not living up to their regulatory responsibilities. The first is "sue-and-settle," a process in which an advocacy group sues a regulatory agency for alleged violations of its responsibilities under the law. Instead of defending itself at a civil or criminal trial, the agency settles with the advocacy group (Tyson, 2014). It is interesting that Scott Pruitt, the Trump administration's first EPA administrator sought to eliminate the sue-and-settle practice. He claimed it was collusion of the agency with environmental groups "behind closed doors." A former career EPA attorney said in his 18 years no case had been settled in court as a result of a suit, and the Biden administration rescinded Pruitt's edict (Brugger, 2022).

A second term used frequently in financing of issue advocacy and election campaigns is "dark money." This is money donated to nonprofit organizations or anonymous corporate entities (e.g., a super PAC) that are politically active. Recipients may spend the money to influence election or other policy campaigns, but they are not required to disclose the source of the money. In the decade since 2010, dark money groups have spent roughly $1 billion—on TV, online ads, mailers, and so on—to influence elections

(OpenSecrets, 2021). A third term, "sustainable communities," references the principle of providing for future generations "as much of and as good as" what the current generation receives. The communities include local (e.g., towns, cities, even university campuses), the subnational unit (a U.S. state), nation-states, and the globe (Rosenbaum, 2020).

Environmental litigation against government agencies. Between 1989 and 2005, the Forest Service (FS) was involved in 949 court cases. The most frequent litigants suing the FS were environmental organizations (12 in all), which is to be expected as the number of laws supporting environmental protection significantly outnumber laws supporting greater resource use. The other litigants were forest industry firms, wood products companies, or other federal government agencies. A surprising finding was the diversity of parties supporting the Forest Service, including environmental groups like the Sierra Club and Wilderness Society, which were willing to spend funds to defend the agency's activities. On some occasions they were major opponents of FS, on others major supporters, based on the type of land management activity and circumstances of the controversies (Portuese et al., 2009).

In 2011, the ranking member of the U.S. Senate Committee on Environment and Public Works, James Inhofe (Okla.-R) and the chair, David Vitter, asked Congress's Governmental Accountability Office (GAO) to determine if there was a trend in the number of cases filed from the late 1990s to 2010 (midway in President Barack Obama's first term). The GAO found that there was no trend in the number of environmental cases brought against EPA. The Department of Justice staff defended EPA on an average of about 155 such cases each year, for a total of 2,500 cases between 1995 and 2010. Most of the cases were filed under the Clean Air Act (59 percent) and the Clean Water Act (20 percent). GAO staff interviewed several stakeholders about these cases. Respondents suggested several factors that affected environmental litigation: change in presidential administration, new regulations or amendments to laws, and failures of the EPA to meet statutory deadlines for actions mandated by Congress (GAO, 2011). These are the most exhaustive and careful studies that have been conducted, and they reveal no trend in litigation.

Changes in environmental group behavior. The litigation strategy was a natural response to the first stage in the evolution of U.S. environmental policy. In the 1960s and 1970s, environmental protection through regulation of industry enjoyed bipartisan support. However, at this early period, the laws and regulations were dictates, examples of "command-and-control," tolerating no negotiation. Penalties and litigation were considered appropriate remedies for failure of compliance. The Reagan presidency changed this ethos as the relationship between industry and environmental groups evolved.

At the beginning of the modern period of environmental law, industry and trade associations employed cost-benefit analysis (CBA) as the basis for environmental standards (and especially to justify any increase in stringency). Environmental NGOs), on the other hand, were opposed to the use of CBA because they believed that some environmental benefits could not be quantified. For example, how do you put a dollar value on the pleasure of seeing an unspoiled mountain lake or paddling down a river through a peaceful forest?

With experience, staff in environmental departments and agencies became more comfortable and adept at using CBA; then some industry groups and their trade associations reversed positions, arguing that CBA should not be used because it could be interpreted to justify stringent regulations. Environmental NGOs reconsidered their positions as well (Livermore & Revesz, 2015).

Starting in the 1980s, progressive business firms turned in the direction of collaboration, crossing economic sector and political party lines in pursuit of environmentally sustainable and responsible ways of doing business. The term applied was the corporate social responsibility (CSR) ethic, directing business firms' attention away from short-term profits and toward longer-term collaborative relations. Polling data in 2019 indicated that a majority of respondents across party lines (including 53 percent of Republican voters) supported the plan to tax companies' greenhouse gas emissions and return the money to taxpayers (Cama, 2019). Environmental groups found this plan much weaker than the carbon tax they supported in the Clinton administration, but it was a gesture favoring collaboration.

When President Trump withdrew the United States from the international Paris Climate Agreement, environmental NGOs formed a broad climate coalition, and networked with social justice groups. Their broad platform encompassed water access, infrastructure and energy-sector emissions, and broad-scaled economic development (Sobczyk, 2019). Environmental NGOs also expressed increased openness to negotiations with the U.S. Chamber of Commerce, as the latter began to take climate change more seriously.

A few U.S. environmental NGOs established ties with European and Asian groups, yet until the last decade, most concentrated exclusively on domestic environmental policy. This began to change in the late 1990s and reached a crescendo at the time of the UN Climate Change (Madrid) Summit in late 2019.

Social media facilitated collaborative efforts and made possible global demonstrations. The technology allowed the spreading of environmental messages rapidly in a dynamic format. NGO members collected information and translated it into personal narratives, sharing them on Facebook and Twitter (Dosemagen, 2017).

Diversity within the environmental NGO community. Ronald Reagan's election to the presidency in 1980 marked a turning point in American attitudes toward agencies, laws, and regulations crafted to protect environmental resources. During the Reagan years, it became conventional wisdom among many conservative and business constituencies that environmentalists were "extreme" and wanted government regulators to stop economic development and curb individual property rights to an unreasonable level. In 2021, most conservative and alternate right perspectives retain this view. To be sure, members of environmental organizations vote for Democrats at a far higher rate than they support Republicans. Still, there are conservative groups such as The Nature Conservancy (TNC) that acquire important funding from corporations and put into practice conservation ideas supported by conservative think tanks such as the Cato Institute and Heritage Foundation [Sobczyk, 2020].

Dark money and environmental NGOs. A final question about environmental NGOs is whether any of them benefit from "dark money" funding sources that are not transparent. Most environmental groups—including the most prominent and influential ones—rely on membership contributions and transparent donations for their operations, but dark money is an income source for a few environmental NGOs. A group called Western Values Project (WVP) became prominent by exposing the close connections to industry and suspected ethics violations of Ryan Zinke, a Montana congressman who became first secretary of the Interior in the Trump administration. Zinke alleged the WVP revelations amounted to a hatchet job from President Trump's critics. Tax records showed that WVP's chief financial backer received support from an electric utility and groups connected to major oil companies, as well as some critics of the Trump administration. The watchdog group Center for Responsive Politics commented: "It's hypocritical to promote transparency as an organization and not have transparency about your finances" (Hiar, 2019).

FURTHER READING

Brugger, Kelsey. "EPA revokes Trump-era 'sue and settle' memo," *Greenwire*, March 24, 2022.

Cama, Timothy. "BP, Shell pledge $1M for carbon tax push," *E&E News*, May 20, 2019.

Dosemagen, Shannon. "Social media and saving the environment: Clicktivism or real change?" *Huffington Post*, January 28, 2017.

Farah, Niina. "Report slams GWU regulatory center for Koch ties," *E&E News*, June 3, 2019.

Hiar, Corbin. "Dark money fuels transparency-focused conservation group," *E&E News*, January 7, 2019.

Iaconangelo, David. "'Numbers don't lie,' Energy sector's diversity questioned," *E&E News*, June 5, 2020.

Jacobs, Jeremy. "Report: Enviro groups diversity—but it's slow going," *E&E News*, January 14, 2021.

Johnson, Stefanie. "Leaking talent: How people of color are pushed out of environmental organizations," Green 2.0, June 2019. https://drstefjohnson.com/leaking-talent-how-people-of-color-are-pushed-out-of-environmental-organizations/

Kraft, Michael. *Environmental Policy and Politics*, 6th ed. New York: Routledge, 2015.

Livermore, Michael and Richard Revesz. "Interest groups and environmental policy: Inconsistent positions and missed opportunities," *Environmental Law*, Vol. 45, no. 1 (Winter 2015): 1181–85.

OpenSecrets. "Dark money basics," 2021. https://www.opensecrets.org/dark-money/basics

Portuese, Gambino, Robert Malmsheimer, Amanda Anderson, Donald Floyd, and Denise Keele. "Litigants' Characteristics and Outcomes in US Forest Service Land-Management Cases 1989 to 2005," *Journal of Forestry*, Vol 107, no. 1 (January/February 2009): 16–22.

Purcell, Kristen and Aaron Smith. "Section 1: The state of groups and voluntary organizations in America," *Internet & Technology*. Pew Research Center, 2011. https://www.pewresearch.org/internet/2011/01/18/section-1-the-state-of-groups-and-voluntary-organizations-in-america/

Rosenbaum, Walter A. *Environmental Politics and Policy*, 11th ed. Washington, DC: Sage/Congressional Quarterly, 2020.

Sobczyk, Nick. "Greens, social justice groups build climate coalition," *E&E News*, July 18, 2019.

Sobczyk, Nick. "Republican operatives launch climate group," *E&E News*, May 6, 2020.

Tyson, Ben. "An empirical analysis of sue-and-settle in environmental litigation," *Virginia Law Review*, Vol. 100, no. 2 (November 2014): 1545–1601.

U.S. Congress, Governmental Accountability Office (GAO). Environmental litigation: Cases against EPA and associated costs over time," August 2011.

Weiss, Haley. "Youth climate strikers: 'Shut down Trump—not the EPA!,'" *E&E News*, September 27, 2019.

Q37. IS THE PROPOSED GREEN NEW DEAL (GND) A FEASIBLE MODEL FOR ENVIRONMENTAL IMPROVEMENT IN THE UNITED STATES?

Answer: Yes, it is feasible, but not likely to be implemented in the near to midterm.

The Facts: The massive legislative package collectively known as the Green New Deal is named after Franklin D. Roosevelt's sweeping "New Deal" economic revitalization programs during the Great Depression. It concentrates both on environmental improvements (reducing CO2 emissions) and social justice (eradication of poverty).

Green New Deal legislation was introduced to Congress in early 2019 by Rep. Alexandria Ocasio-Cortez (D-NY) and Sen. Ed Markey (D-Mass), but the idea of revitalizing American infrastructure and economy with sweeping new investments in green energy is older. One source is a *New York Times* article in 2010 by journalist Thomas Friedman; the second is Green Party candidates in state/federal elections since 2010 whose platforms emphasized energy sustainability and combating climate change.

The Green New Deal is both a legislative proposal and a movement. One group that has lined up in support of the legislation, for example, is the Sunrise Movement, a grassroots effort to combat climate change that seeks to harness the idealism and energy of young people. Political director Evan Weber presented its ideals: "These insurgent campaigns are a clear indicator of the appetite for an entire new way of doing things, and a restructuring of our society under a populist agenda that guarantees things like living wage jobs, affordable and safe housing, universal clean air and water, and Medicare for All—all policies which we see bundled into the Green New Deal framework" (Cama, 2019).

Decarbonization. This term connotes complete elimination of carbon-based energy products in industry, the commercial sector, government, and in people's homes (Northey, 2018). An initial proposal to eliminate fossil fuels in 10 years was made less extreme by setting a target of 2050 instead. Second, a carbon pricing proposal was revived, as an incentive for business firms and consumers to use fewer fossil fuels and to create revenue for return to taxpayers or to reduce the federal budget deficit. A 2019 report of the Congressional Research Service (CRS) indicated that potential benefits of a carbon tax would depend on "the program's magnitude and design and most importantly the use of carbon tax revenues" (CRS, 2019).

Most supporters of renewables recognized that a natural gas bridge from fossil fuels to clean energy would be needed for many years.

Smart grid. A frequently mentioned option to improve the national electricity grid (see chapter 4) is to construct a network of long-distance, ultra-high-voltage transmission lines, which would widely share wind and solar power across the different time zones of the continental United States. Surplus solar power of the Southwest in the afternoon could stream into Southeastern states at sundown. Other lines could transmit unused wind energy from the Midwest to the Great Lakes and East Coast regions (Willrich, 2017). DOE's National Renewable Energy Laboratory (NREL) estimated that benefits of such a grid would be three times greater than the cost. Why, then, has there been opposition? One factor is the NIMBY (not-in-my-backyard) phenomenon, in which Americans who are supportive in theory of certain types of investment—such as turbines for wind energy—express objections to having those investments made in their particular communities. A second factor is competition and conflict among the states within a region; a third is a politically polarized Congress that is often deadlocked when it comes to making substantive changes to environmental or energy policy.

Peter Taft, an engineer of grid transformation in DOE's Pacific Northwest National Laboratory, criticized the GND for failing to account for the profound operating changes (especially increased speed and complexity) of the U.S. electric power sector in the last century: "A future with potentially 30 percent of the U.S. installed resource capacity coming from distributed resources and customer participation requires a different physical distribution system than exists today" (Behr, 2019). Since Taft made these remarks in 2016, no large infrastructure project to redesign the grid has been developed.

Costs. Most opposition to the Green New Deal focuses on the price tag of implementing all its ambitious proposals. The actual cost of GMD is difficult to discern. One back-of-the-envelope estimate is $93 trillion, which is greater than four times the value of the U.S. GDP in 2021, but this figure includes proposals for sweeping new investments in social welfare programs in addition to the "green" policies for which the Green New Deal is named (Klein, 2019).

The Green New Deal and the 2020 elections. Most of the Democratic Party's presidential candidates expressed support for the broad contours of the GND, but the party nominee, Joe Biden, was lukewarm and pondered labor's objections. America's largest unions in general have supported campaigns fighting climate change, but as AFL-CIO president Richard Trumka said about GND: "We weren't part of the process, and so the worker's interest really wasn't completely figured into it. . . . So we would want a whole lot of changes made so that workers and our jobs are protected in the process" (Sobczyk, 2019a).

Republicans, meanwhile, saw the Green New Deal as an opportunity to paint their Democratic opponents as antibusiness and fiscally irresponsible. The politically conservative, Republican Party-aligned American Action Forum was one such group opposed to the Green New Deal. A year before the 2020 presidential election, the Forum's affiliate, the American Action Network, surveyed voters in 30 congressional districts where the two parties were competitive. Respondents saw the progressive climate platform of GND favorably—48 percent to 46 percent, when pollsters described it as a plan "to address climate change and income inequality, and transition the United States from an economy built on fossil fuels to one driven by clean energy." When respondents heard that GND might cost $93 trillion, and were told that "the plan would not stop climate change because countries like China would not make similar changes," respondents opposed it 61 percent to 32 percent (Sobczyk, 2019b). These results show the impact that the wording of poll questions can have on responses.

A transformational Biden cabinet? Although Donald Trump never conceded his loss in the 2020 election, Biden won both popular and electoral vote majorities conclusively. The Biden administration's focus on climate change as a high-priority issue was reflected in the appointment of former presidential candidate (and retired Senator) John Kerry as climate ambassador, and former EPA administrator Gina McCarthy as White House "national climate advisor." The key energy and environment positions went to experienced leaders from state government (such as former Michigan governor Jennifer Granholm as Energy secretary; Michael Regan, environment conservation chief of North Carolina, as EPA administrator; and Debra Haaland, former member of Congress and tribal leader, as Interior secretary.

Infrastructure spending in the Biden administration. In 2021, the Biden administration engineered the passage of a bipartisan $1 trillion infrastructure package that made major federal investments in climate resilience initiatives, transportation (including electric vehicles and Amtrak), the national power grid, and roads, bridges, and airports.

Special Senate rules allow essential budget legislation to avoid a filibuster and pass the chamber with just a majority. President Biden proposed a $3.5 trillion social spending package, which was approved in a nonbinding budget resolution, parceled out to relevant committees. Only mandatory, previously authorized expenditures, such as Social Security, could be included. Making consensus difficult were disagreements between two conservative senators, Joe Manchin of West Virginia and Kyrsten Sinema of Arizona, and the other 48 senators of the caucus, and between the House GND caucus of 96 members (not necessarily reflective of their stance on all GND issues) and moderate members of the House and Senate. (As the 2022 midterm elections approached, neither party was optimistic that a majority could be reached for any package [Cama, 2022].)

After several months of negotiations, President Biden announced a $1.75 trillion framework to which most agreed. Removed from the package were ambitious climate change programs, provisions to provide federal paid leave, a large expansion of Medicare, and two years of free community college. Remaining in the package were universal preschool for more than 6 million 3- and 4-year-old children and subsidies to limit costs of childcare to no more than 7 percent of income for most families. The largest investment in attacking climate change was $555 billion in incentives for Americans to purchase EVs and for utilities to move from natural gas and coal to renewables (Weisman et al., 2021). The bill was the largest infrastructure spending package passed by Congress in more than a decade. Although the 10-year total was $1 trillion, the new federal spending was $550 billion (with an estimated increase in the deficit of $256 billion).

The Senate vote in favor of the bill was 69-30, which was unusually bipartisan. Ignoring shrill efforts of former president Trump to kill it, Senate minority leader Mitch McConnell said he was "proud to support it." To gain support of sufficient Republican votes, however, Biden and his Democratic allies in Congress had to give up switching out lead pipes, as well as additional transit and clean energy projects. They also had to accept "pork barrel" projects benefitting districts/states not based on national need, such as reconstruction of an Alaska highway and restoration of the San Francisco and Chesapeake bays (Cochrane, 2021).

FURTHER READING

Behr, Peter. "Power lines: The next 'Green New Deal' battlefront?" *E&E News*, February 25, 2019.

Burnett, Sara. "Progressives say primary wins latest sign of momentum shift," *U.S. News*, August 5, 2020.

Cahlink, George. "Trump, House GOP agree on key to 2020 win: Green New Deal," *E&E News*, September 13, 2019.

Cama, Timothy. "Sunrise Movement looks to unseat slate of Democrats," *E&E News*, December 12, 2019.

Cama, Timothy. " 'Travesty'; Left and right slam spending bill," *Greenwire*, March 9, 2022.

Cochrane, Emily. "Senate passes $1 trillion infrastructure bill, handing Biden a bipartisan win," *New York Times*, August 10, 2021.

Huber, Matt. "Why the Green New Deal has failed—so far," *Jacobin Magazine*, May 2021. https://jacobinmag.com/2021/05/green-new-deal-climate-change

Jordan, Larry. *The Green New Deal: Why We Need It and Can't Live Without It . . . and No, It's Not Socialism.* Coppell, TX: PageTurner Books Intl., 2019.

Klein, Naomi. *On Fire: The (Burning) Case for a Green New Deal.* New York: Simon & Schuster, 2019.

Northey, Hannah. "What exactly is the 'Green New Deal'?" *E&E News*, November 16, 2018

Shaffrey, Mary. "AP calls N.Y. race against Engel; Maloney still undecided," *E&E News*, July 17, 2020.

Sobczyk, Nick. "Union chief says no to Green New Deal," *E&E News*, April 24, 2019a.

Sobczyk, Nick. "Poll says Green New Deal unpopular—as described by the GOP," *E&E News*, August 14, 2019b.

U.S. Congress, Congressional Research Service (CRS). "Attaching a price to greenhouse gas emissions with a carbon tax or emissions fee: Considerations and potential impacts," March 22, 2019. https://fas.org/sgp/crs/misc/R45625.pdf

Waldman, Scott. "Trump warns against Green New Deal: 'It's like baby talk,'" *E&E News*, June 5, 2020.

Weisman, Jonathan, Jim Tankersley, and Emily Cochrane. "Biden implores Democrats to embrace $1.85 trillion climate and safety net plan," *New York Times*, October 28, 2021.

Willrich, Mason. *Modernizing America's Electric Infrastructure.* Cambridge, MA: MIT Press, 2017.

7

Conclusions

This volume is driven by questions, and as originally designed, only one issue seemed apt for the concluding chapter: the "energy transition"—the movement from fossil fuel energy sources to renewables. The advent of both an economic recession and a public health pandemic in early 2020 changed this calculation, as they influenced all questions in the book to some extent. Questions surrounding the nature and extent of the recovery of the American and world economies still abound in early 2022 (when this book was completed), as do questions about the pandemic and its ultimate impact on public health, law, and society.

This book concludes by considering two important energy development questions deeply intertwined with present economic and public health challenges. Q38 touches on coverage in chapters 1 through 3 to consider whether the energy transition from fossil fuels to renewables has been stopped or delayed by the crises that spread around the world in 2020, 2021, and early 2022. Few commentators believe that either the pandemic or economic volatility will return America to an era of reliance on fossil fuels predominantly; the relevant issue is whether in the next generation, renewables will be uppermost. Question 39 summarizes many observations of chapters 4 through 6. It asks whether separation of powers (especially the courts' increasing strength) and federalism have become major determinants of U.S. energy and environmental outcomes, at times of extreme partisan polarization. Has the increased activism of the conservative Supreme Court increased overall uncertainty, reducing momentum for adoption of renewables? Although some discussion reaches back to the Obama administration, most attention is paid to the last two years of the Trump administration and the first 15 months of the Biden presidency.

Q38. HAS THE U.S. ENERGY TRANSITION FROM FOSSIL FUELS TO RENEWABLES BEEN STOPPED OR MERELY DELAYED BY ECONOMIC VOLATILITY AND THE COVID-19 GLOBAL HEALTH CRISIS?

Answer: During the Trump administration, the energy transition slowed. It quickened in the first year of the Biden administration and then paused for several reasons treated in this question.

The Facts: Popular environmental theorist Lester Brown examines the shift from fossil fuels to clean energy, remarking that this monumental transition "will compress a half-century of change into the next decade" (Brown et al., 2015). This was a reasonable calculation when Brown completed his book, just five years before the outbreak of the COVID-19 pandemic. Now it seems less realistic.

Discussion on the question has five parts, starting with fossil fuels (oil, natural gas, and coal), comprising well over two-thirds (79 percent) of U.S. energy production and consumption in 2020 (EIA, 2021b). These are primary sources of energy in the United States, and recently they have been subject to immense volatility because of both market instability and the pandemic. The second section considers the smaller part of the U.S. energy profile, nuclear and hydropower. and then wind, solar, and other renewables. The third section treats the electricity grid and electric vehicles (EVs). The fourth section looks at governments' defense of these vital energy and transportation sectors, and the massive stimulus programs developed to mitigate the crises. The last section takes stock of overarching trends in both fossil fuels and green energy alternatives.

Fossil fuels (oil, natural gas, coal). In late 2019 and early 2020, a price war between Saudi Arabia and Russia for control of the global oil market pushed prices to extremely low levels. Because American shale oil producers had higher costs than oil producers in most Middle Eastern countries, they were hard hit. The Energy Information Agency (EIA) forecast that oil production would not rise to pre-COVID levels until sometime in 2022 (EIA, 2021a). Eruption of the COVID-19 pandemic in early 2020 and its cascading effects seriously compounded the downward spiral of oil prices, which dropped below zero briefly at the end of the first quarter. Drilling rigs sat idle, layoffs spread, and the number of storage places for surplus oil vanished quickly (Clark & Lee, 2020).

As the oil market collapsed and financing for production declined, shale oil and gas producers cut costs and production—but not in time to ward off an oil and gas glut. While most drillers were burning off (flaring) surplus gas, some drillers in Texas pumped excess gas down wells (Cunningham, 2019). The size of firms engaged in conventional oil/gas E&P is larger, and they were better able to outlast price volatility. Liquefied natural gas (LNG) export prospects dimmed considerably, and natural gas prices dropped because of "ravaged global oil demand" and a supply surplus. In the major shale gas/oil concentrations of the Permian, Eagle Ford, and Bakken basins, new fracking operations dropped nearly 70 percent from March to April 2020. However, in 2022 crude oil prices had rebounded and reached the highest level since October 2014 ($120/barrel for West Texas Intermediate). President Biden and EU leaders imposed economic sanctions on Russia, the globe's third-largest supplier of oil and natural gas, when it invaded the Ukraine (Richards, 2022). As economic growth rebounded in OECD nations, prices increased sharply due to pent-up demand for most commodities. Also, concerns had eased about the impact of the omicron strain (a very infectious variant of COVID-19) and new unrest in the Mideast.

Coal's decline was rapid too, but it rallied to a level below oil. From January 2017 to January 2021, the Trump administration compiled a consistent pro-coal record from the perspective of the industry, whereas the Obama administration had a mixed record (and the Biden administration showed a clear preference to further reduce coal usage in its first year [Chemnick et al., 2022]). Efforts of different administrations, however, had less of an impact on the coal industry than market forces, as utilities and other energy consumers turned away from coal, citing its greater expense as well as adverse environmental consequences of coal use ranging from polluted water to climate change. As a result, coal—which from 1950 to 2010 accounted for 50 percent or greater of the electricity produced in the United States—appears destined to become a minor source of energy. Moody's Investors Service reported that coal would comprise just 11 percent of U.S. power generation by 2030, as compared to the 2019 rate of about one-quarter of all power (Gladstone, 2019).

Hydropower, nuclear, and other renewables. Hydro- and nuclear power contributed 7 and 20 percent, respectively, to U.S. electricity production and smaller parts of the national energy mix in 2020. (Nuclear is not a renewable, but its fossil source [uranium] is less quickly exhaustible than that of the other fossil fuels.) Significant obstacles remain to making either one a larger part of the U.S. energy profile. For hydropower, problems include aging infrastructure of America's large dams, a confusing pattern of ownership, and adverse environmental effects. In addition, a lot of the best sites for large-scale hydropower projects in the United States have long since been developed. Problems facing an expansion of nuclear power are even greater. Not only are existing nuclear power plants old, but nuclear waste disposal is an unresolved issue, dividing the public. Substantial government and private-sector investment would be needed to increase both nuclear and hydropower. The closure rate of nuclear power plants increasingly resembles that of coal power plants. Both hydro- and nuclear power are more attractive prospective energy sources when designed as smaller plants. Strides have been made in planning and development of micro-reactors, especially for use in military installations and for rural areas distant from the grid.

Turning to wind, the United States was the second-largest wind power market in the world, and on average wind accounted for 9 percent of U.S. electricity generation in 2020 (EIA, 2021). Solar energy enjoyed a boom in 2020 as well, accounting for 3 percent of U.S. electricity needs. Increasing economies of scale and declining materials' costs have made solar and wind much more competitive with fossil fuels. Yet slowing economic growth and state and community mobility restrictions had adverse effects. The American wind industry developed later than those in China and Europe, and it remains highly dependent on a global supply chain. Factory shutdowns because of the two crises disrupted import of raw materials. Also, offshore wind development required construction permits and support from financially stressed Northeast coast state governments.

The situation of solar is comparable, with the exception that funding solar power projects has relied on the investment tax credit (ITC), the value of which has declined since its enactment in 2006. Described by the Solar Energy Industries Association (SEIA) as "one of the most important federal policy mechanisms to incentivize clean energy in the United States," the ITC still requires annual congressional appropriations (Ferris, 2020b).

The energy industry lost about 15 percent of its jobs because of the COVID-19 pandemic. Wind and solar lost relatively more jobs than fossil fuels because many workers sustained cuts in hours or wages (IER, 2020). The largest losses occurred in energy efficiency companies, whose work more resembled restaurants than manufacturing plants. Employees conducted efficiency upgrades at homes and businesses; making personal visits and shaking hands with homeowners and proprietors were difficult in the pandemic. Many firms closed soon after states and cities adopted social distancing rules and travel restrictions (Ferris, 2020a).

Challenges to wind and solar power development included acquiring sufficient raw materials and storage capacity, infrastructure needs, investment gaps, and social and environmental obstacles. Were these obstacles overcome, the renewables still would not compete in price and general prospects with natural gas, even as natural gas prices rose slowly in 2021. The incoming Biden administration's embrace of a strong climate change agenda appeared to spur wind and solar prospects greatly. However, advocacy pursuing "zero emissions by 2050" encountered resistance from the fossil fuel industry, the transportation industry, and the Republican party at multiple scales. In early 2022, the base of the Republican party remained loyal to former President Trump and unwilling to compromise on measures that would reduce America's dependence on fossil fuels.

Finally, the coronavirus affected the biofuel industry, because oil prices dropped so low (for a time) that oil and gas were cheaper than biofuels. Travel restrictions also reduced sales of gasoline. However, in 2020, farmers increased corn acreage over 2019 (one reason being that farmers wanted more corn to feed surplus livestock they were not able to sell abroad or at home). By 2021, the USDA Economic Research Service (ERS) was forecasting a record corn yield for the 2021/22 corn crop. Use of corn for ethanol was expected to increase by 5 percent, and use of corn for seed, animal feed, and industrial purposes to increase by 4 percent (ERS, 2021).

The electricity grid and EVs. America's energy grid, which included 3,000 utilities in 2017, is composed of five distinct and interconnected regional grids. The system includes two major grids, one in the East (including the area east of the Rocky Mountains and part of the Texas panhandle) and one in the West (from the Rockies to the Pacific). It also includes three minor grids: the Electric Reliability Council of Texas (ERCOT) including most of Texas; the Alaska grid; and the Canadian power grid (EIA, 2021). (The Canadian grid is integrated into the U.S. grid, which allows U.S. regions [New England states, New York, California, Upper Midwest, and Pacific Northwest states] to receive electricity supplies from the network. More than 35 transmission interconnections link the U.S. and Canadian systems.)

The sources of electricity in the United States are diverse, as are the technologies used to generate them. In 2020, natural gas was the largest source (about 40 percent) of total electricity generation. It was used to fire steam and gas turbines to make electricity. Coal was the country's third-largest source of electricity generation in 2020, accounting for about 19 percent. Most of the power plants burning coal used steam turbines; a few converted coal to gas for use in gas turbines. Less than 1 percent of electricity was generated from petroleum (residual fuel oil, petroleum coke, or diesel). Nuclear energy provided some 20 percent of the U.S. electricity profile in 2020. The nuclear power plants used steam turbines, producing power from nuclear fission. The breakdown of renewables' 20 percent was: wind (8.4 percent), hydro (7.3 percent), solar (2.3 percent), biomass such as wood (1.4 percent), and geothermal (0.4 percent) (EIA, 2021).

The grid system for which most information is available is the New York Independent System Operator (NYISO). Initial reports suggested a slight decline in usage when the pandemic hit the state, as well as a significant "unprecedented" shift in the times of highest energy demand due to office closures, lockdowns, and people staying at home telecommuting. After Governor Andrew Cuomo issued a statewide lockdown directive, utilities maintained 24 hours of operation without interruption (Northey & Iaconangelo, 2020). Directives (first from state governors and then President Trump) banned utilities from shutting off electricity and water for those unable to pay their bills.

A 2021 BloombergNEF study on sustainable energy use in the United States found that the pandemic caused a significant plunge in energy use. All sources of energy (fossil fuels and renewables) dropped by 7.8 percent in 2020, the largest drop in 30 years' federal record-keeping. Car and truck fuels experienced the largest decline; electricity use fell by 3.8 percent because office buildings and industrial facilities were closed or heating/cooling systems were used less. Renewables (solar, wind, hydropower) filled the gap by adding record amounts of electricity to the grid (Ferris, 2021). Economists at Brattle, an energy consulting group, assessed the impact: "The utilities' cost of capital likely has increased due to increased volatility and cost-recovery risks" associated with demand reductions from social distancing and consumer anxiety (and delayed utilities' bill payments) (The Brattle Group, 2020).

Despite its aging infrastructure and a relatively high number of outages, the U.S. power grid meets routine expectations. Most transformers and transmission lines are a quarter-century old or more; the average age of power plants is 34. The country has twice as many power plants as needed because of the inefficiencies built into the system. This has the effect of increasing outages, which grow annually. The United States has the greatest number of outage minutes of any economically developed country. Extreme outages (blackouts) occurred in California (August 2020) and Texas (February 2021).

After deregulation of the California electricity system in 2000, rolling blackouts allowed the old infrastructure to operate, but lax maintenance of the state's largest utility, Pacific Gas & Electric Co. (PG&E), was a factor blamed for forest fires in 2017/2018, leading to bankruptcy of the utility. California also endured inadequate electricity supplies during the

August 2020 heat wave (Mulkern & Behr, 2020). Another kind of extreme weather condition—unprecedented cold temperatures and ice—caused major electricity outages in Texas and other states in 2021. The wintry weather killed 200; 4 million homes and businesses were without power; and 12 million residents lacked access to clean drinking water. Energy researcher Joshua Reynolds said, "It's such a black swan event. . . . It is taxing every single piece of the system at the same time" (Klump et al., 2021). Lack of grid reliability and resilience were driving forces behind the Obama administration's 2015 unveiling of a multiyear Grid Modernization Initiative (GMI). Crafted by the Department of Energy, the GMI was an effort to develop a cost-effective roadmap for a reliable, secure, and sustainable energy grid that would still be affordable, both for the nation and individual consumers (see Q20, ch. 4).

Similar problems of sustainability affected electric vehicles (EVs) during the recession and start of the pandemic. The drop in oil prices (and in turn, gasoline and diesel fuels) briefly removed a major advantage of EVs, their lower potential operating costs. Also, cautious car buyers saw electric cars as risky, given inaccessibility of charging stations and limited battery storage. An additional risk factor was the greater dependence of EVs on the global supply chain (and especially Chinese production of cobalt and lithium), as trade conflicts of U.S. and China did not lessen. Yet BloombergNEF argued that the "tried-and-true" automobiles would fare well during times of crisis and disruption, but that "the long-term electrification of transport is projected to accelerate in the years ahead" (Ferris, 2020c).

At the start of the pro-oil Trump administration, prospects for EVs were not rosy. In the last two years of the administration, however, the outlook seemed to brighten. EV makers such as Elon Musk demonstrated success at constructing larger batteries. Battery prices subsequently dropped, making EV production more competitive with gas guzzlers. Only 1 of 65 EV models planned was canceled. At the start of the Biden administration, with its strong focus on clean energy and climate change, it would appear that the fortunes for electric vehicles are trending much more positively.

COVID-19 impacts on energy and electricity sectors. Responses occurred in four stages: mobilization of the nation, organization of governments to identify the virus/develop vaccines and disseminate them, federal monetary policy (Federal Reserve Bank action), and fiscal policy (stimulus spending involving the president and Congress). The Trump and Biden administrations figured differently in each stage.

The Trump administration was criticized for its failure to mobilize and develop effective countermeasures in the pandemic's early months, which in turn resulted in tens of thousands of deaths (Sumner, 2020). Confusion as to whether the president should use emergency powers to curb disease transmission or if state governors should act immediately (as they had clear health and welfare powers under their constitutions) delayed this response, as did shortages of diagnostic testing and medical supplies. Vaccine development began in January 2020 and involved government vaccine researchers in collaboration with pharmaceutical, biotech, and academic partners. These efforts resulted in the NIH-Moderna and Pfizer/BioNTech vaccines, produced and authorized by the Food & Drug Administration (FDA) for use in the United States within a matter of months [Bok et al., 2021].)

By early March 2020, Trump had issued a "guidance document" restricting nonessential travel, setting social distancing and outdoor gathering size, and closing nonessential businesses (such as restaurants, movie theatres, etc.). Most state governors quickly declared emergencies and issued detailed restrictions on individual mobility and business activity, but several did not do so—an early signal of partisan division that worsened as the pandemic spread. This lack of national clarity, in the judgment of experts, increased risk of exposure and unnecessary deaths. In response, the Trump administration announced a partnership among several federal agencies to speed up development, manufacturing, and distribution of COVID-19 vaccines.

Operation Warp Speed, unveiled in April 2020, gave the Trump administration a platform and some needed muscle. When medical equipment, particularly ventilators, seemed likely to be insufficient, the president exercised emergency provisions of the Defense Production Act to reconfigure factories to produce them. Senior administrators also determined which industries were "critical," including telecommunications, defense, agriculture, transportation and logistics, electric power, petroleum, water, and ultimately even the coal industry.

The "critical" designation gave guidance on which sectors and industries were "essential" to keep the country running, despite pandemic-related restrictions for critical infrastructure operations—from defense, to health care, and agriculture. Critical or essential workers continued to work during periods of community restrictions, social distancing, or closure orders; nonessential workers were to follow state or federal lockdown orders and stay at home. Second, the designation indicated a priority scheme for distribution of the vaccines in January 2021. In addition to the distinction between vulnerable persons (based on age, underlying medical conditions, residence in long-term care facilities) and others, essential workers such as those in health care, first responders, even coal miners had access to earliest doses of vaccines available.

Former vice president Joe Biden was a 2020 presidential candidate when the pandemic unfolded. Like every other Democratic candidate, he heavily criticized Trump's slow response to the crisis, his contradictory statements about the pandemic, and his politicization of public health decisions.

By contrast, Biden sought greater transparency in communications with the public about the pandemic and vaccination efforts. He also returned the Centers for Disease Control (CDC) to a position of independence from the White House, notwithstanding frustration with the agency's withholding critical data on the administration's booster shot plan (Banco et al., 2021). Among his first actions as president was to issue a national strategy for the COVID-19 Response and Pandemic Preparedness; his immediate goal was that 70 percent of all adult Americans have at least one vaccination by July 4, 2021; the administration fell just short of this goal (reaching 67.1 percent) due to strong resistance to vaccination, especially in so-called red states. At this time the Delta variant became the dominant COVID-19 strain in the United States, indicating it was highly unlikely that "herd immunity" could be attained, given the creative and adaptive nature of the virus. Thus, the mission changed from eradication to mitigation.

After 10 years of strong economic growth, the oil market crash in late 2019, and the onset of COVID-19 in early 2020, unemployment rates soared. In 2019, the average rate of unemployed was 3.67 percent. By the second quarter of 2020, it had reached 8.31 percent, the largest quarterly increase in the previous 30 years (the number of unemployed in the second quarter was 20.6 million, much higher than the peak of 15.1 million in the Great Depression). The unemployment rate in October 2021, however, fell to 4.6 percent (BLS, 2021).

As part of executive actions taken during his first day in office, Biden continued some projects of his predecessor. He extended federal student loan forgiveness and extended the moratorium on evictions. The first large ($1.9 trillion) stimulus and relief package of the Biden administration was the American Rescue Plan Act of 2021, which extended unemployment benefits and made direct payments to individuals. In addition, it had several tax provisions, such as expanding the child tax credit and child/dependent care credits. More remarkable than this largely partisan package was the $1.2 trillion Infrastructure Investment and Jobs Act, one of the largest infrastructure bills to pass Congress in American history. About $550 billion was in new spending, including new roads and bridges (comprehending transportation safety programs; public transit; passenger and freight rails; EV infrastructure, electric buses; airports, ports, and waterways; resilience and Western water infrastructure; broadband Internet; and environmental remediation (Cama, 2021). The total spending for stimulus and COVID-19 relief totaled more than $7 trillion. After passage of the infrastructure bill, support for further broad-scale spending declined precipitously. Republican legislators opposed these efforts strongly; not only conservative outliers in the Democratic party but also moderates could not countenance support for popular items such as energy tax credits and resumption of child/dependent care credits when the 8.5 percent inflation rate in early months of 2022 was the highest rate since the early 1980s.

The changing face of American energy. During the 2016 presidential campaign, Donald Trump said he would end the "war on American energy,"— by which he meant hostility to fossil fuels—and three years into his presidency he claimed that the United States had attained "energy dominance." The Trump administration played a broker role seeking reduction in global oil production from OPEC countries and Russia as oil prices went negative. The president tweeted: "We will never let the great U.S. Oil & Gas Industry down. . . . I have instructed the Secretary of Energy and Secretary of the Treasury to formulate a plan . . . so that these very important companies and jobs will be secured long into the future" (Dillon, 2020). But while Trump gave full-throated support to oil, gas, and coal, he maintained his hostility to the wind and solar energy sectors.

In his first address to Congress, President Biden announced a "blue-collar blueprint" to change the fossil fuel mindset of Americans. He used the speech to emphasize the need to rebuild the nation's transportation and energy sectors in order to address the climate crisis, while also creating jobs for American workers. The speech was not a "renewables dominant" argument, although he spoke of the great potential for development of wind, solar, and other non-carbon energy sources. He also spoke of advantages in using green energy to power the American electricity system, another source of good jobs (Clark, 2021).

Unlike many renewables advocates, President Biden sees a continued role in the American energy mix for natural gas. The Eastern Interconnection Planning Collaborative, a nonprofit whose members operate the high-voltage power grid across the Eastern and central American states, recommended that policymakers consider a "reliability safety valve" allowing utilities more time to meet wind and solar power goals by using natural gas as a bridge.

Fossil fuel interests and advocates of renewables use the same terms when they discuss the energy transition, but as *New York Times* reporters Gelles and Friedman point out, they mean different things. Supporters of wind, solar, and nuclear say the transition calls for speedy elimination of fossil fuels; petroleum industry executives believe the transmission means continued use of fossil fuels and perhaps replacement of coal with natural gas (Gelles & Friedman, 2022). That's why in the near- and midterm, the transition will be bumpy.

FURTHER READING

Banco, Erin, Sarah Owermohle, and Adam Cancryn. "Tensions mount between CDC and Biden health team over boosters," *Politico*, September 13, 2021.

Bok, Karin, Sandra Sitar, Barney Graham, and John Mascola. "Accelerated COVID-19 vaccine development: Milestones, lessons, and prospects," *Immunity*, Vol. 54, no. 8 (August 10, 2021): 1636–51.

The Brattle Group. "Brattle economists publish assessment on COVID-19 impacts on energy industry." April 14, 2020. https://www.brattle.com /insights-events/publications/brattle-economists-publish-assessment-on -covid-19-impacts-on-energy-industry/

Brown, Lester, Janet Larsen, J. Matthew Roney, and Emily Adams. *The Great Transition: Shifting from Fossil Fuels to Solar and Wind Energy*. New York: W. W. Norton, 2015.

Cama, Timothy. "How the infrastructure bill happened and what it will do," *E&E News*, November 8, 2021.

Chemnick, Jean, Hannah Northey, and Sean Reilly. "Biden preps full-court press to curb coal as emissions spike," *E&E News*, January 21, 2022.

Clark, Lesley. "Biden lays out 'blue-collar blueprint' to transform energy," *E&E News*, April 29, 2021.

Clark, Lesley and Mike Lee. "Trump: 'You're going to lose an industry,'" *E&E News*, April 1, 2020.

Cunningham, Nick. "Oilfield services face crisis as shale slowdown worsens," *Oil Price*, October 21, 2019. https://oilprice.com/Energy/Energy -General/...

Department of Commerce, Bureau of Labor Statistics (BLS). "The Employment Situation—October 2021," https://www.bls.gov/news.release/archives /empsit_11052021...

Dillon, Jeremy. "Trump wants lifeline for struggling industry," *E&E News*, April 21, 2020.

Ferris, David. "Inside clean energy's coronavirus job crash," *E&E News*, April 8, 2020a.

Ferris, David. "Solar industry to lose over a third of jobs by July," *E&E News*, May 18, 2020b.

Ferris, David. "Analysis: Crisis will spark 'electrification of transport,'" *E&E News*, May 19, 2020c.

Ferris, David. "Pandemic causes biggest plunge in energy use in 30 years," *E&E News*, February 18, 2021.

Gelles, David and Lisa Friedman. "There's a Messaging Battle Right Now Over America's Energy Future," *New York Times*, March 20, 2022.

Gladstone, Alexander. "Moody's downgrades coal sector on weakening export demand," *Wall Street Journal*, August 21, 2019.

Institute for Energy Research (IER). "Analysis inflates clean energy job losses due to pandemic," July 24, 2020. https://www.instituteforenergyresearch.org

Klump, Edward, Lesley Clark, and Mike Lee. "'Heads will roll': Grid crisis sparks political firestorm," *E&E News*, February 22, 2021.

Lee, Mike. "Oil CEO: 'We will disappear. . . like the coal industry,'" *E&E News*, April 15, 2020.

Mulkern, Anne and Peter Behr. "Blackouts threaten heat-ravaged grid," *Energywire*, August 18, 2020.

Mulkern, Anne. "What Calif. blackouts reveal about U.S. grid," *E&E News*, March 16, 2021.

Northey, Hannah and David Iaconangelo. "'We have never done this before.' Inside N.Y.'s grid lockdown," *E&E News*, March 30, 2020.

Richards, Heather. "Biden bans Russian oil, backing Ukraine," *Greenwire*, March 8, 2022.

Sumner, Mark. "Stunning visualization shows how many lives could've been saved if Donald Trump had acted sooner," *Daily Kos*, April 15, 2020.

U.S. Department of Agriculture (USDA), Economic Research Service (ERS). "Market outlook," February 19, 2021. https://www.ers.usda.gov/topics/crops/corn-and-other-feedgrains/market-outlook

U.S. Department of Energy, Energy Information Administration (EIA). "U.S. oil and natural gas production to fall in 2021, then rise in 2022," January 14, 2021a. https://www.eia.gov/todayineergy/detail.php?id=46476

U.S. Department of Energy, Energy Information Administration (EIA). "U.S. energy facts explained," May 14, 2021b. https://www.eia.gov/energyexplained/us-energy-facts/

Wells, Ester. "EVs to reach 10% of all vehicle sales for first time—report," *E&E News*, November 9, 2021.

Q39. HAVE THE SEPARATION OF POWERS (ESPECIALLY THE COURTS' INCREASING STRENGTH) AND FEDERALISM BECOME MAJOR DETERMINANTS OF U.S. ENERGY AND ENVIRONMENTAL OUTCOMES?

Answer: Yes. At a time of extreme partisan polarization, they rival economic factors (and cultural forces) in influence.

The Facts: The framers of the U.S. Constitution believed that "auxiliary precautions" were essential to constrain the drive of popular leaders, and for that reason they established the separation of powers system, with three coequal branches of government—the executive branch (the presidency), the legislative branch (Congress), and the judicial branch (the courts). They also believed that the system of states would protect the people's liberty from potential tyranny of a powerful national government. This question considers whether these safeguards work in setting energy and environmental policy outcomes in the twenty-first century.

There have been several episodes of presidents seeking to expand their powers since the FDR administration of the 1930s–40s. During these periods, institutional checks by the Congress and Supreme Court often restrained impulses of popular leaders. A second institution, the federal bureaucracy, is part of the executive branch, and specialized for rulemaking. Bureaucracies are pressured by the White House, Congress, the courts, and also political parties, industry, the media, and nonprofits (e.g., environmental groups and foundations). When presidents prioritize, as both presidents Trump and Biden did in three departments/agencies, chances of policy success are great (Potter & Shipan, 2019). Finally, the White House and state governors have competed and cooperated with respect to the direction of energy and environmental policy. This pattern of "dual federalism," present since 1789, often is conflict-laden.

Courts' blocking of Trump administration actions. In 1835, Alexis de Tocqueville famously observed: "Scarcely any political question arises in the United States that is not resolved, sooner or later, into a judicial question" (Mansfield & Winthrop, 2000). Clear evidence of the courts' continued ability to restrain the executive were decisions in high-stakes environmental issues. For instance, the majority (6-3) opinion in County of Maui v. Hawaii Wildlife Fund steered between a major expansion of the Clean Water Act and allowing regulated entities (such as wastewater plants) to avoid permitting requirements (King, 2020). From 2017 through 2020, the Trump administration deregulated elements of the Obama administration energy and environmental policy legacy (two cases are discussed below). Courts deferred Trump administration rollbacks for consideration by the Biden administration.

Congressional curbs on Trump administration budget action. The Constitution gives Congress primary budget authority, which it exercises through its powers of authorization and appropriation. Since the Great Depression, presidents have whittled away at this authority, and President Trump was no exception. In deliberations on the FY 2020 budget (beginning on October 1, 2019), members of appropriations committees in the House and Senate quickly rejected deep cuts proposed by the White House to several EPA, Interior, and Energy programs. They paused when considering a new proposal to move the Bureau of Land Management (BLM) headquarters from Washington, D.C., to Grand Junction, Colorado. The rationale given was that most U.S. public lands were in the Western states, but the likely reason was the need to "drain the swamp," the topic of Q26. Neither body approved funding the Interior Department's management move (Lumney & Bogardus, 2019), and an early action of the Biden administration canceled it.

Bicameralism and checks on Biden administration appointments. Even with token Democratic control of the Senate, it remained an obstacle in the first year of the Biden administration. The Constitution gives the Senate the power to "advise and consent" to presidential appointments, which is a method used by the opposition party to delay or impede the president's agenda. Opposition to Biden appointees was based on the sharply different direction taken by the new administration—in climate warming, clean energy, endangered species protection, and in its attempts to diversify the upper reaches of the bureaucracy with respect to race, ethnicity, gender, and sexual orientation. Partnership for Public Service, a nonprofit, nonpartisan research institute, evaluated the speed in confirmation of the presidents from George W. Bush to Biden. It found that 144 of Biden's nominees were confirmed compared to 130 of Trump's nominees, but 291 nominees of George W. Bush and 304 of Obama (Bogardus, 2021). Increased polarization of politics explains much of the difference, and outreach to the other party had grown more difficult. Moreover, in the Biden administration, intraparty differences (especially between progressives and moderates) were intractable.

Congressional limitation on Biden administration budget actions. The Biden administration had a bold fiscal policy agenda including huge spending bills, for example, the $1.2 trillion infrastructure bill. In the Senate it had bipartisan sponsors, as the Biden administration made changes in its initial proposal to gain some Republican support (such as ensuring there was some pork in every state's barrel). Again, minorities within majorities exercised influence. For instance, to gain support of West Virginia Senator Joe Manchin, a $150 billion program to assist utilities in transitioning from fossil fuels was deleted; to gain support of Arizona Senator Kyrsten Sinema, the proposal to reverse Trump tax rate reductions for high earners and corporations was rejected (Sobczyk, 2021).

Court restrictions on Biden executive action. Some actions of U.S. district courts, such as pausing several rollbacks of major environmental statutes, supported the incoming Biden administration. A few cases, however, frustrated the new president. The first case was the Clean Power Plan (CPP), referred to in Q31; the Supreme Court agreed to hear appeals from Republican-led states and coal companies to limit the EPA's authority to regulate carbon emissions under the Clean Air Act. The nearly equal party division in Congress meant that Supreme Court reluctance to defer to the agency or the incoming Biden administration would stop federal attempts to reduce greenhouse gas emissions (Liptak, 2021; see also King, 2022). A second case was the Waters of the U.S. (WOTUS) rule (determination of which waterways and wetlands are automatically protected under the Clean Water Act [CWA]). The Supreme Court has agreed to ascertain whether a lower court correctly handled a challenge from Iowa landowners attempting to build on their own land without procuring an expensive federal permit, and a positive court finding would narrow the orbit of the CWA (Northey, 2022).

Federal-state conflict on environmental issues. The federal government requires the states to protect the environment and mitigate externalities such as pollution in broad legislative mandates. Conflict may arise both when the federal regulation establishes a floor (a mandate states may not want to have put in place) or a ceiling (a limitation on additional action a state may want to take). Review of the Clean Air Act (CAA) shows the ways in which state-level action constrained policy objectives of the Trump administration (and, to a lesser extent, the Biden presidency). The CAA regulates air emissions from stationary and mobile sources; it authorizes EPA to establish standards to protect public health and welfare and to regulate emissions of hazardous air pollutants. Applying CAA in 2019, the EPA administrator Wheeler accused the California Air Resources Board (CARB) of having "the worst air quality in the United States." The CARB director, Mary Nichols, reminded Wheeler that "CARB was established years before U.S. EPA came into existence" (Joselow, 2019). In this case, the Biden administration moved quickly to restore the authority to California, because it fit well in the climate change and clean energy strategy of the new presidency.

The second case pit the states of Michigan, Minnesota, and North and South Dakota against the Federal Energy Regulatory Commission (FERC). The states enacted laws allowing sole-source bids in construction of new large electricity transmission lines in their service areas, while FERC has a regulation (Order 1000) opening them up to competition. The idea behind the FERC regulation was that competition would stimulate new transmission projects and lower the cost of the projects (facilitating development of a national electricity grid). But utilities in the states lobbied state legislatures to pass right-of-first-refusal laws, which increased tension between state and federal energy policy (Tomich, 2022).

Summary observations. The United States has a high degree of partisan polarization in the twenty-first century; moderates have been pushed to the edge (or squeezed out) of both the Democratic and Republican parties due to gerrymandering and other factors, and the divisiveness has extended deeper into the electorate than in recent times (as noted by declining numbers of independents). Waning of bipartisanship is correlated with political gridlock, and this is a poor climate for resolution of issues through national legislation. Neither party has held both the White House and Congress for a long period since the administration of Franklin D. Roosevelt, and frequent power reversals produce a seesaw effect (and uncertainty) for groups and interests directly affected by energy and environmental policy.

Presidents with ambitious agendas often attempt to circumvent some of this gridlock by using executive orders and instructing their agencies to develop new policies and regulations that do not need approval from Congress. Opponents know that federal courts will not reliably defer to executive orders and regulations, and increasingly they are inclined to file suits against agencies. Perhaps the most significant accomplishment of Trump's presidency was securing the confirmation of three conservative judges to the Supreme Court, giving conservatives a 6-3 majority. With a clear conservative majority, the Court has shown an inclination to side with the interests of established energy industries (oil/gas and coal industries and their trade associations) rather than with the interest groups representing renewables and the environment. (As noted in Chapter 6, the Supreme Court has made increasingly frequent use of the shadow docket, especially on issues in which the minority's position is likely to draw heightened public interest [Vladek, 2022].)

The Supreme Court within the last three years (2019 to 2022) has intervened in energy and environmental areas, as cases concerning the CAA, CPP, and CWA demonstrate. However, unlike both the executive and legislative branches of the U.S. government, the judiciary makes decisions on the briefs presented by others: it does not possess specialized knowledge of its own; there is no "environmental" justice on the Supreme Court.

The Biden administration has, as of early 2022, emphasized a "comprehensive strategy" for energy that attempts to sidestep the courts when possible. Yet activist judges and justices have the capacity to create uncertainty for those seeking stability in energy and environmental policy. This book shows that environmental issues and energy security risks increase the value and importance of renewable energy. It is now generally considered a key factor in reducing greenhouse gas emissions and thus climate warming. However, many observers express concern that conservative activism on the bench could hinder the adoption of renewable energy options seen as essential in combating climate change.

FURTHER READING

Bogardus, Kevin. "Biden's EPA noms face sluggish confirmation," *E&E News*, August 19, 2021.

Joselow, Maxine. "Calif. air regulator fires back at Wheeler," *E&E News*, April 29, 2019.

King, Pamela. "Roberts' court finds the middle in high-stakes enviro term," *E&E News*, July 13, 2020.

King, Pamela. "Inside a legal doctrine that could derail Biden climate regs," *Greenwire*, April 11, 2022.

King, Pamela and Hannah Northey. "Inside Stephen Breyer's Clean Water Act legacy," *E&E News*, January 28, 2022.

Liptak, Adam. "Supreme Court to hear case on E.P.A.'s power to limit carbon emissions," *New York Times*, October 29, 2021.

Lumney, Kellie and Kevin Bogardus. "Senate unveils interior-EPA bill, rejects money for BLM move," *E&E News*, September 24, 2019.

Mansfield, Harvey and Debra Winthrop, eds. *Alexis de Tocqueville's: Democracy in America*. Chicago: University of Chicago Press, 2000.

Northey, Hannah. "EPA rule in the crosshairs as Clean Water Act heads to court," *E&E News*, January 25, 2022.

Potter, Rachel and Charles Shipan. "Agency rulemaking in a separation of powers system," *Journal of Public Policy*, Vol. 39, no. 1 (March 2019): 89–113.

Sobczyk, Nick. "House passes $1.7T climate change, social spending package," *E&E News*, November 19, 2021.

Tomich, Jeffrey. "States unwind FERC plans for grid expansion," *E&E News*, January 19, 2022.

Vladeck, Stephen. "Roberts Has Lost Control of the Supreme Court," *New York Times*, April 13, 2022.

Abbreviations

5G	Five gigabyte
AAF	American Action Forum
ACCF	American Council on Capital Formation
ACE	Affordable Clean Energy program (of Trump administration)
ACES	Advanced Clean Energy Storage (project name)
ACP	American Clean Power Association
AEP	American Electric Power Co.
AFL-CIO	American Federation of Labor-Congress of Industrial Organizations
AFPM	American Fuel and Petrochemical Manufacturers
AGI	American Geosciences Institute
ALEC	American Legislative Exchange Council
AMLER	Abandoned Mine Land Economic Revitalization
ANILCA	Alaska National Interest Lands Conservation Act (of 1980)
ANWR	Arctic National Wildlife Refuge
AP	Associated Press
APA	Administrative Procedures Act
API	American Petroleum Institute
APPA	American Public Power Association
APS	Arizona Public Service
ARPA-E	Advanced Research Projects Agency-Energy
ASRC	Arctic Slope Regional Corporation
AWEA	American Wind Energy Association

bbl	Barrels
BIWF	Block Island Wind Farm
BLM	Bureau of Land Management (DOI)
BNEF	Bloomberg New Energy Finance (media organization)
BOEM	Bureau of Ocean Energy Management (DOI)
BP	British Petroleum
BPA	Bonneville Power Authority
BPS	Bulk power system
BSEE	Bureau of Safety and Environmental Enforcement (DOI)
C3	Conservative Coalition for Climate Solutions
CAA	Clean Air Act of 1970, amended in 1990
CAFE	Corporate average fleet economy (standards)
CAP	Center for American Progress
CARB	California Air Resources Board
CARES	Coronavirus Aid, Relief and Economic Security Act, 2020
CASAC	Clean Air Scientific Advisory Committee
CBA	Cost-benefit analysis
CBD	Center for Biological Diversity
CCC	Civilian Conservation Corps
CEA	Council of Economic Advisors (White House)
CEO	Chief executive officer
CEQ	Council on Environmental Quality (White House)
CERCLA	Comprehensive Environmental Response, Compensation, and Liability Act
CFO	Chief financial officer
CFR	Code of Federal Regulations
CIA	Central Intelligence Agency
CICC	Critical Infrastructure Command Center
CO2	Carbon dioxide
CoCC	Commodity Credit Corporation
COE	Corps of Engineers (U.S. Army)
CoFR	Council on Foreign Relations
COVID-19	Corona Virus number 19
CPP	Clean Power Plan (of Obama administration)
CRS	Congressional Research Service (U.S. Library of Congress)
CSE	Citizens for a Sound Economy
CSIS	Cybersecurity and Infrastructure Security Agency (DHS)
CSP	Concentrating solar power
CSR	Corporate Social Responsibility

CWA	Clean Water Act, 1972
CWP	Center for Western Priorities
CZMA	Coastal Zone Management Act
DARPA	Defense Advanced Research Products Administration (DOD)
DHS	Department of Homeland Security
DLR	Dynamic line rating
DNI	Director of National Intelligence
DOC	Department of Commerce
DOD	Department of Defense
DOE	Department of Energy
DOI	Department of Interior
DOJ	Department of Justice
DOT	Department of Transportation
DRECP	Desert Renewable Energy Conservation Plan
E15	Ethanol (quotient of fuel is) 15 percent
E&E News	*Energy & Environment News*
EA	Environmental assessment
EDF	Environmental Defense Fund
EDGI	Environmental Data & Governance Initiative
EEI	Edison Electric Institute
EERE	Energy Efficiency & Renewable Energy, Office of (DOE)
EGS	Enhanced geothermal system
EIA	Energy Information Administration (DOE)
EIS	Environmental impact statement
EISA	Energy Independence and Security Act, 2007
EJ	Environmental Justice plan (Biden administration)
EMP	Electromagnetic pulses
EO	Executive order (of the U.S. president)
EPA	Environmental Protection Administration
EPRI	Electric Power Research Institute
ERCOT	Electric Reliability Council of Texas
ESA	Endangered Species Act (of 1973)
EVs	Electric vehicles
EWG	Environment Working Group
FACA	Federal Advisory Committee Act
FAST	Fixing America's Surface Transportation Act, 2015

FCC	Federal Communications Commission
FCSC	Federal Cybersecurity Commission
FDA	Food & Drug Association (DHHS)
FDR	Franklin Delano Roosevelt
FEMA	Federal Emergency Management Administration (DHS)
FERC	Federal Energy Regulatory Commission
FFVs	Flexible-fueled vehicles
FLPMA	Federal Land Policy and Management Act, 1976
FOIA	Freedom of Information-Act
FPC	Federal Power Commission
FS	Forest Service (USDA)
FVRA	Federal Vacancies Reform Act
FWS	Fish and Wildlife Service (DOI)
GAO	Government Accountability Office
GDP	Gross domestic product
GHGs	Greenhouse gases
GM	General Motors
GMD	Geomagnetic disturbances
GMI	Grid Modernization Initiative
GND	Green New Deal
GW	Gigawatts
HF	Hydraulic fracturing
ICC	Interstate Commerce Commission
IEA	International Energy Agency
IG	Inspector general
IMF	International Monetary Fund
IPAA	Independent Petroleum Association of America
IPCC	Intergovernmental Panel on Climate Change
IQA	Information Quality Act
IT	Information technology
ITC	Investment tax credit
ITIF	Information Technology & Innovation Foundation
LCV	League of Conservation Voters
LED	Light-emitting diode
LIHEAP	Low income home energy assistance program
LNG	Liquefied natural gas

MBI	Market-based initiatives
MISO	Midcontinent Independent System Operator
mpg	Miles per gallon
MTBE	Methyl tert-butyl ether
MW	Megawatts
NAAQS	National Ambient Air Quality Standards
NAERM	North American Energy Resilience Model
NAM	National Association of Manufacturers
NARUC	National Association of Regulatory Utility Commissioners
NASEM	National Academies of Sciences, Engineering, and Medicine
NATO	North Atlantic Treaty Organization
NCSL	National Conference of State Legislatures
NEPA	National Environmental Policy Act, 1969
NERC	North American Electric Reliability Corporation
NETL	National Energy Technology Laboratory
NGC	National Gas Council
NGOs	Nongovernmental organizations
NHTSA	National Highway Traffic and Safety Administration
NIABY	Not in any backyard
NIAC	National Infrastructure Advisory Council
NIMBY	Not in my backyard
NMA	National Mining Association
NOAA	National Oceanic and Atmospheric Administration (DOC)
NORC	National Opinion Research Center
NOx	Nitrous oxides
NPDES	National Pollution Discharge Elimination System
NPPD	National Protection and Program Directorate
NRC	Nuclear Regulatory Commission
NRDC	Natural Resources Defense Council
NREL	National Renewable Energy Laboratory (DOE)
NWPR	Navigable Waters Protection Rule (of Trump administration)
NYISO	New York Independent System Operator
OCS	Outer Continental Shelf
OIRA	Office of Information and Regulatory Affairs (White House)
OMB	Office of Management & Budget (White House)
OPEC	Organization of the Petroleum Exporting Countries
OPM	Office of Personnel Management (White House)

OSMIRE	Office of Surface Mining Reclamation and Enforcement (DOI)
OSTP	Office of Science & Technology Policy (White House)
OT	Operating technology
PAC	Political action committee
PEER	Public Employees for Environmental Responsibility
PFAS	Per- and polyfluoroalkyl substances
PFCs	Perfluorinated chemicals
PG&E	Pacific Gas & Electric Co.
PHMSA	Pipeline and Hazardous Materials Safety Administration (DOT)
PJM	Pennsylvania-New Jersey-Maryland Interconnections Ltd.
PL	Public law
PM	Particulate matter
PR	Public relations
PSH	Pumped storage hydropower plants
PUCs	Public utility commissions
QTR	Quadrennial technology review
R&D	Research and development
RADICs	Rapid Attack Detection, Isolation and Characterization Systems
REEs	Rare-earth elements
RFA	Renewable Fuels Association
RFF	Resources for the Future
RFG	Reformulated gasoline
RFS	Renewable fuel standard
RTOs	Regional transportation organizations
SAB	Scientific Advisory Boards
SAFER	Safer Affordable Fuel Efficient Vehicles Rule (of Trump administration)
SBA	Small Business Administration
SDWA	Safe Drinking Water Act, 1974
SEIA	Solar Energy Industries Association
SMRs	Small Modular Reactors
SOP	Separation of powers
SPR	Strategic Petroleum Reserve
SSI	Supplemental security income

TANF	Temporary assistance for needy families
TAPS	Trans-Alaska Pipeline System
TMI	Three Mile Island
TNC	The Nature Conservancy
TRC	Texas Railroad Commission
TSA	Transportation Security Administration
TURN	The Utility Reform Network
TVA	Tennessee Valley Authority
TXOGA	Texas Oil & Gas Association
UCS	Union of Concerned Scientists
UIC	Underground injection control
UNCED	United Nations Conference on Environment and Development
USBR	U.S. Bureau of Reclamation (DOI)
USDA	U.S. Department of Agriculture
USGS	U.S. Geological Survey (DOI)
VEETC	Volumetric Ethanol Excise Tax Credit
W&WPTO	Water and Water Power Technologies Office (DOI)
WECC	Western Electricity Coordinating Council
WOTUS	Waters of the United States (of Obama administration)
WVP	Western Values Project

Index

About the Author

Jerry A. McBeath is professor of political science emeritus, University of Alaska Fairbanks. His research and publications span energy and environmental policy and politics (U.S. and international); political economy (U.S., China, Taiwan); and Arctic studies (including Alaska specialization).